CONTENTS

CHAPTER ONE
The Bayer Post Intro

Winter slams into Aspen earlier and runs off later than it does in the civilized world. While most other general contractors were warming up their front-loaders in anticipation of an early ground-breaking, I was bidding Spring remodels while hoping for an odd-ball restoration project or custom-designed whatever to make life interesting and pay the rent. New construction was *happening* in 1999, but being the jack-of-all-trades that I was, I gravitated more toward small, odd-ball projects that were construction-related, and paid equally well, but involved ingenuity more than man-power. I preferred commissions to build custom furniture, cabinets, and sculpture in wood, metal, stone, and whatever. As luck would have it, that project came along in the form of an invitation to bid on the reproduction of a wood sculpture titled *Totem* by Herbert Bayer.

Herbert Bayer was one of the founding fathers of the Aspen Institute and the man responsible for bringing the Bauhaus school of art to America in the 1940's. Having restored several of his other sculptures in concrete and metal for the Aspen Institute it was assumed I could reproduce his *Totem*, an incredible and still relatively unrecognized example of modern wood sculpture. The Totem shown below is a subsequent reproduction of it and it's still featured on my website, www.falbergsawz.com. I consider it a gross injustice to Herbert Bayer that what I consider his best work is virtually unknown to the art world. But I'm neither an artist nor an art expert. I'm just a fan and I enjoy visiting such objects of interest. My visit with *Totem* was especially interesting and ultimately changed my life forever.

Herbert Bayer's *Totem* is a vertical post of Douglas Fir twelve feet tall that has been standing, exposed to the elements, since 1952, in the front yard of his former residence in Aspen. It consists of symmetrical semi-circles as the reciprocating segments of two wavy rip-cuts along the length of a 14" x 14" x 12' timber. The originally exterior edges, when inverted, formed the center-line of the finished piece. The exposed radii of its linked, intersecting, semicircular cuts expose two distinct patterns which oppose each other linearly. It takes my breath away just saying that! I'll provide more pictures and a detailed layout of the Bayer post in chapter eight for anyone who's interested in the details. At the time it looked like an impossible task and my inquisitive nature absolutely relished the challenge of it.

My client, myself, and the Curator of Denver's esteemed Art Museum, Gwen Chanzit, drove up the mountain to see it and take some pictures. It was both an interesting puzzle to reverse engineer and a challenging project to execute. That it needed saving, for the sake of posterity, was immediately obvious from the weathered gray of sun-exposure on its exposed surfaces, the black decay of water stains under its shaded coves, and the sporadic coloring left by the peeling remnants of its original clear-coat finish. To my eye, and judging by its grain, it was made from a fine specimen of old-growth Douglas Fir. We talked briefly of the deceased artists copyrights but saw no reason not to reproduce it, if we could.

Later that day I sat for hours staring at Polaroids of the sculpture and getting dizzy with the mental effort of understanding it. It took some studying but eventually I recognized the pattern of laterally reciprocating radii being the result of two rip cuts in a single 10" x 10" timber. I know this version of the technical explanation can't be the correct way of describing Bayer's design because it changes every time I try to put it into so few words but it's the best I can do with my limited writing ability. No matter how I say it, however, you still wouldn't understand without seeing the pictures. The pictures of the original piece were so bad, being

Polaroid, that I now show the second reproduction instead:

Fig. 1-02

Understanding the design, in this instance, was not the same as knowing how to do it; so I visited the Aspen library hoping to find some background on how it was done. The only book I could find relating to the *Totem* project showed the two German guys credited with its construction standing in front of a mammoth behemoth bandsaw. There was no *Totem* in the picture nor a beam from which to make it. They both had big smiles for the camera. The article accompanying the photo claimed only that it was done by these two, without explaining how. The reader was left to assume that *Totem* was accomplished by maneuvering said Douglas Fir beam through a long series of perfectly accurate semi-circles, on the saw pictured, by some well co-ordinated miracle of slow-motion shop choreography that would have had to be flawlessly executed to get the result I saw in Bayer's yard. I didn't believe it then, and I don't believe it now. My conclusion, then and now, was that they'd cut it from four separate pieces and glued them together afterwards just to satisfy Bayer's insane geometric nightmare. I think they did it the hard way, using four smaller posts; but that's still saying something, because the layout must have been intense. I could be wrong. We'll never know. If they're not dead, they're really, really old.

Regardless of how they actually did cut it, (the grain of the old-growth original was so consistent and dense I couldn't prove my theory forensically) the only method I could see to duplicate it was to find a way to make a bandsaw portable enough to perfectly execute those same snaky cuts in real, tough wood. There were several German brands of portable bandsaws available at the time, but unlike the *totem*'s original builders, the saws didn't work very well and none of them had throats wide enough to follow the curves' link-points, in the middle of the beam, where they cross over and start the next semicircle. I'd heard all the stories about guys that put casters on the legs of their bandsaw and rolled the saw around the timbers. That wasn't an option for me, because my timber had a twist and my floor was far from flat or smooth. That's another of those great ideas that look great on paper but totally fall apart in practice.

The only way I could see to get two perfect cuts in this situation was to turn a band saw upside-down and make it portable. Through some impossible cosmic accident there just happened to be a beat-up old Delta three-wheeler, with enough throat width to do the job, for sale at a nearby thrift store; for cheap. It took a radical make-over of that old *anchor* to turn its 250 pound mass into a portable band saw. It didn't matter that I knew nothing about band saws. Nobody made the kind of portable band saw I needed for this cut anyway and I saw no reason to learn the current state-of-the-art, because it wouldn't apply to me. I had to start with the basic concept of a blade in an endless loop transported by two or more wheels. Further definition of a band saw depends on the purpose for which it's to be used and there are too many variations to describe them all here. I was interested in vertical wood-working band saws and more specifically, portable vertical woodworking band saws; of which there were but two that I knew of at the time and both were two-wheelers with narrow throats. Neither served my needs.

Using abrasive cutting wheels in ways their creators never intended, I managed to chop the Delta 28-560 down to its bare-bones-minimum frame elements and strip it down to 150 lean, mean pounds of brute blade-transporting iron. Talk about hot-rodding a tool! It just so happened that I had been involved in a faux-timber framing project the month before all this, and the memory of using a very expensive off-the-shelf German portable band saw was still fresh in my mind. It was one of those two-man operations that depended on the dancing skills of both participants and made me seriously question their claim to *portability*. I've never danced well with other men and we went through twenty blades making four cuts with it. I remembered thinking: had the saw been balanced over the timber, we wouldn't have needed two of us to do it. Balancing the saw over the blade was therefore a top priority for *my* version of a usable portable bandsaw. The most aggravating thing about the German portables was how they persistently leaned to the frame side. And I hate listening to the screaming noise of a universal motor when I'm trying to rock and roll.

Portable being a relative term, I had to rig a block-and-tackle to get my battleship gray monstrosity upside-down on my timber. I still remember it well. It was late in the day at that point and I was still experiencing post-traumatic-stress from cutting cast iron with an exposed abrasive blade at 10,000 RPM (without the safety guard); so I thought I'd run the saw into one end of the beam a little bit, just to get it stabilized on my saw horses for the night. As I commenced to cut I was amazed at how easy it was working and kept going, and going; and before I realized what was happening, I was cutting out the other end. Wow! Done! In, like, ten minutes!

Still unbelieving, I had to roll it over a quarter turn and try again with the remaining two pieces clamped together in their original position. I ran my little test cut again with ever more confidence. And that was it. Done! I had bid the job for $6000 and finished it in twenty minutes. Well, not quite; I still had to glue the straight edges together, finish, and install it. But still; the hardest part was done in the blink of a dusty eyelid. I partied that night and slept with visions of grandeur. Invention mania is a powerful thing! There was no "E*ureka!"* moment. It was more like "Wow, that actually worked?!" It was a lame-brained experiment born of desperation that, by pure accident, worked the first time around. By pure luck I had invented something that worked.

First thing the following morning I set my prized saw on a cheap Chinese utility box to get some good magazine-quality Polaroids of my latest greatest in natural sunlight. I was so excited that I didn't notice the box melting in the heat and when the plastic turned to putty my *Precious* plopped on the concrete floor; breaking its spine. That morning's headline should have read: "Chinese saboteurs cripple American innovation!" It didn't matter. The idea was born. The seed was planted. It was alive!

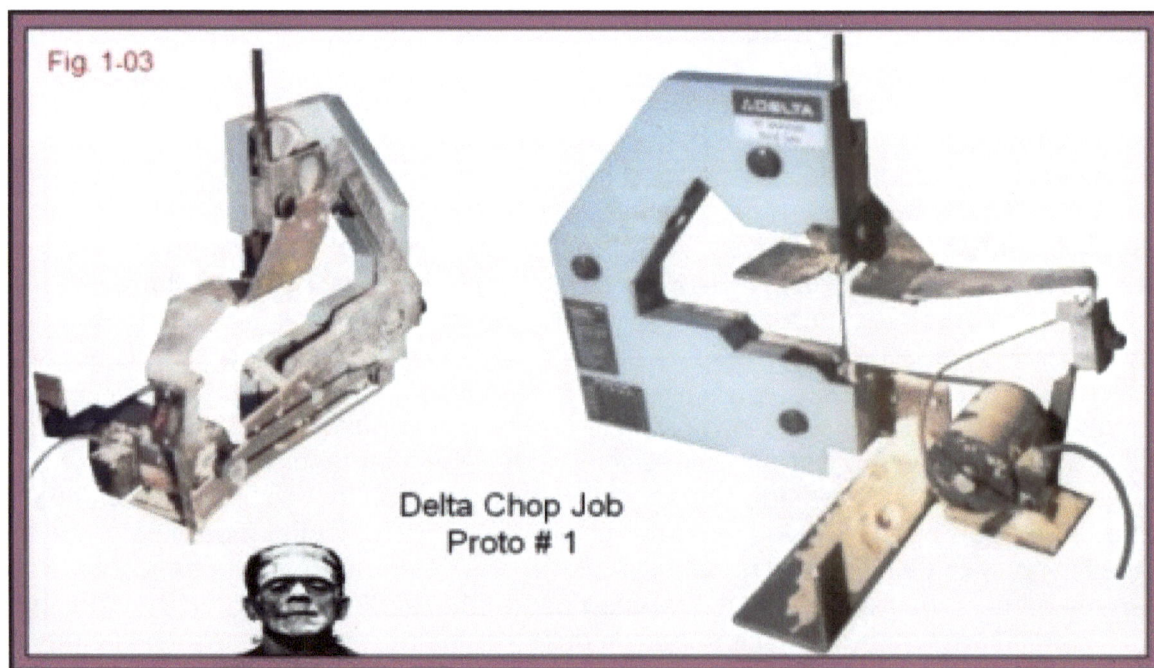

Fig. 1-03

Delta Chop Job
Proto # 1

No stranger to the invention game, and being already the owner of one patent, I had also unsuccessfully pursued two others so I knew how the process worked. I set out immediately therefore to protect the balanced portable band saw concept by filing a provisional patent application and set out to build an aluminum prototype forthwith.

The only band saw I'd ever owned now lay broken beyond repair on my concrete floor. I knew nothing about band saws at that time except that they were very heavy and mostly useless for anything practical, like contracting. My tools were primarily oriented toward general construction; not timber framing or metal working. Being mechanically handy most of my life, however, I had an assortment of tools and a general idea of how things worked; not to mention a healthy heapin' of good old American ingenuity going for me. All I knew for sure was that my prototype had to be balanced, light enough to man-handle, and have a whopping 18"W throat to be the perfectly portable band saw.

Following the same design philosophy I had adopted way long ago, when I wasn't capable of building anything even close to a machine so complicated as a bandsaw, I broke every function down to its minimum requirements. In this case, all it had to do was cut a Bayer post. It didn't have to be quiet, economical, ergonomic, symmetrical, or beautiful. The fewer, the smaller, and the lighter the parts; the better. The first consideration I gave to the weight factor was to stipulate aluminum for its frame. It did not have to be elegant or slick. "Keep it simple", I thought; "just make something that does this one specific job: Bayer posts." There would no doubt be other applications for the idea but those would be for the market to determine after I'd found the throat-size limit of a portable bandsaw.

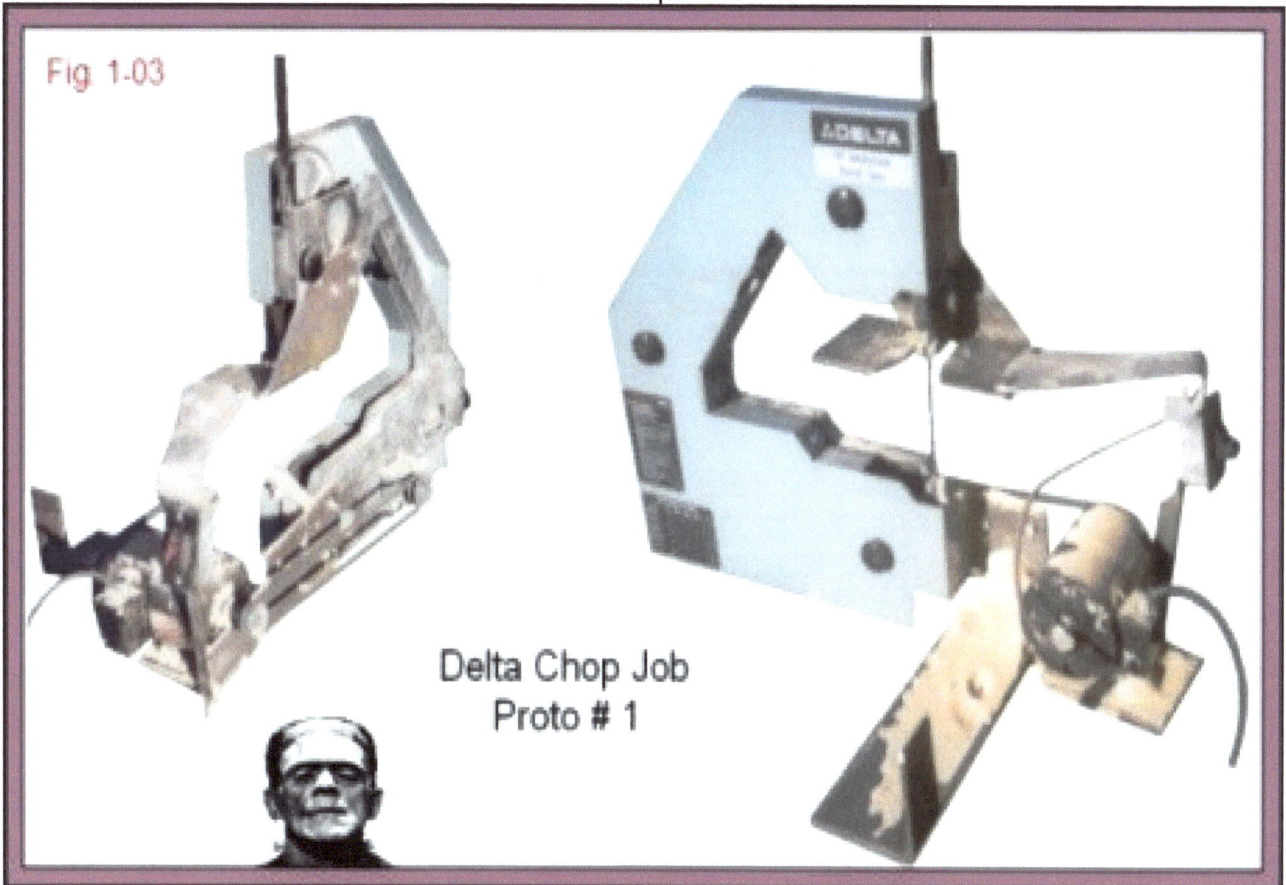

Fig 1-03

Delta Chop Job
Proto # 1

With those simple parameters in mind there arose, after a month or two, prototype saw number one. Still using the original Delta wheels, guides, and ½ HP motor; it featured what might be the first wrap-around throat ever seen in bandsawdom. It was a donut-shaped affair with 16" of width on either side of the blade and 12" of cut depth. But it was balanced and worked like a charm. It sold immediately and the buyer got behind the idea of patenting and producing the saws commercially.

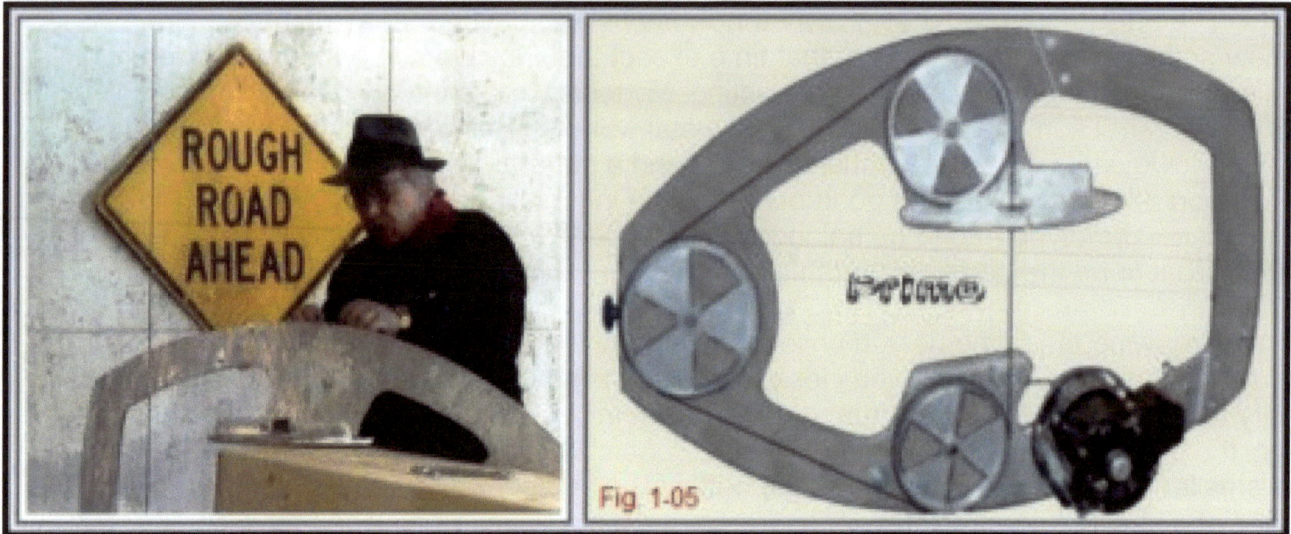

Fig. 1-05

Anyone who has ever designed and built anything is aware of mankind's propensity to over-build; it's the prevalent philosophy of conservative design that says "Better safe than sorry." My challenge was to build the most rigid frame possible with the *least* amount of weight. In a world where band saws were judged by the weight of their frames and wheels, I was going to have to go the other way and change the reality as well as the perception. Conventional wisdom on the subject was not going to apply to me in my quest for a bigger portable band saw.

In order to make a portable band saw with three times the throat area of a big cast-iron two-wheeler, and that out-performed one as well, I would have to rethink the whole concept of band sawing. To me, somehow, 900 lbs of iron just didn't seem necessary to drive a blade that's only 1/2" wide and .032" thick, which happens to be the ideal blade spec for cutting corbels out of timber. New as I was to band sawing, I wasn't even aware that there was conventional wisdom regarding band saws.

That turned out to be a great example of ignorance being bliss. It wasn't immediately obvious that I would have to find devious, outside-the-box alternatives to every tenet of conventional band saw thinking to make these saws, but I could see from the preponderance of cast iron frames and wheels that I would be *swimming upstream* and deliberately avoided any reading material on the subject until after I had built the first prototype of the balanced portable saw. Conventional wisdom would have had me using the biggest motor and wheels I could find. I eventually found that contemporary band saws are, for the most part, designed around out-dated technology, old-wives tales, and myths. Prototypes number two and four were heavier variations of number one, and are still in operation. I knew the throats would eventually have to be open but I was clueless in how to design such frames with sufficient strength. Every saw since then has had an open throat but with varying degrees of rigidity. It took four years of trial and error to design a three-wheeled band saw frame that performed 8" deep cuts like two-wheeled saws performed 6" cuts. At a tenth of their weight. Most of that effort would center on building a strong, but lightweight, frame.

FALBERG
SAW
11

Fig 1.06

12" X 18"
1/4" PLATE FRAME
3/4HP ODP MOTOR
8" DELTA WHEELS
DELTA GUIDES
SINGLE-SPRING
SCREW TENSIONER
POWDER COATED

"The Racing Version"

Saws #5 through #30 were plagued by the tracking problems associated with weak frames and I ended up replacing most of them, free of charge, or bracing whenever possible with an additional cross-brace of aluminum angle. It would be two more years before I discovered the more complex skeletal frame configuration of Model C in which 2" welded tubing was cross-braced by 2" x 2" x 3/16" angle, and making both components serve a dual role. Those two years of trial-and-error experimentation also gave me time to master the metal-working skills I'd theretofore lacked. I ended up making a rather complex assembly jig to fabricate frames of that size and it has been in use ever since.

The Titan Model D, seen below, is the present embodiment of the Falberg portable band saw and will, no doubt, undergo further improvements as we continue to evolve. It measures 18.5" from the platen to the lower blade guide and 18.25" from the blade to the vertical frame member. It's not as aggressive as a 25 HP log mill but it will cut arched timbers from 18" beams all day and it's faster than a chain saw. The dual-spring lever-action tensioner, an un-patented Falberg original design, has been standard equipment since 2003, as are the simple but elegant blade guides and lower extension. At $4900 it is the most expensive portable bandsaw on the market today but the least expensive in terms of dollars per square inch of throat capacity. Not one ounce is squandered nor one dollar spared in the pursuit of lean-ness and mean-ness.

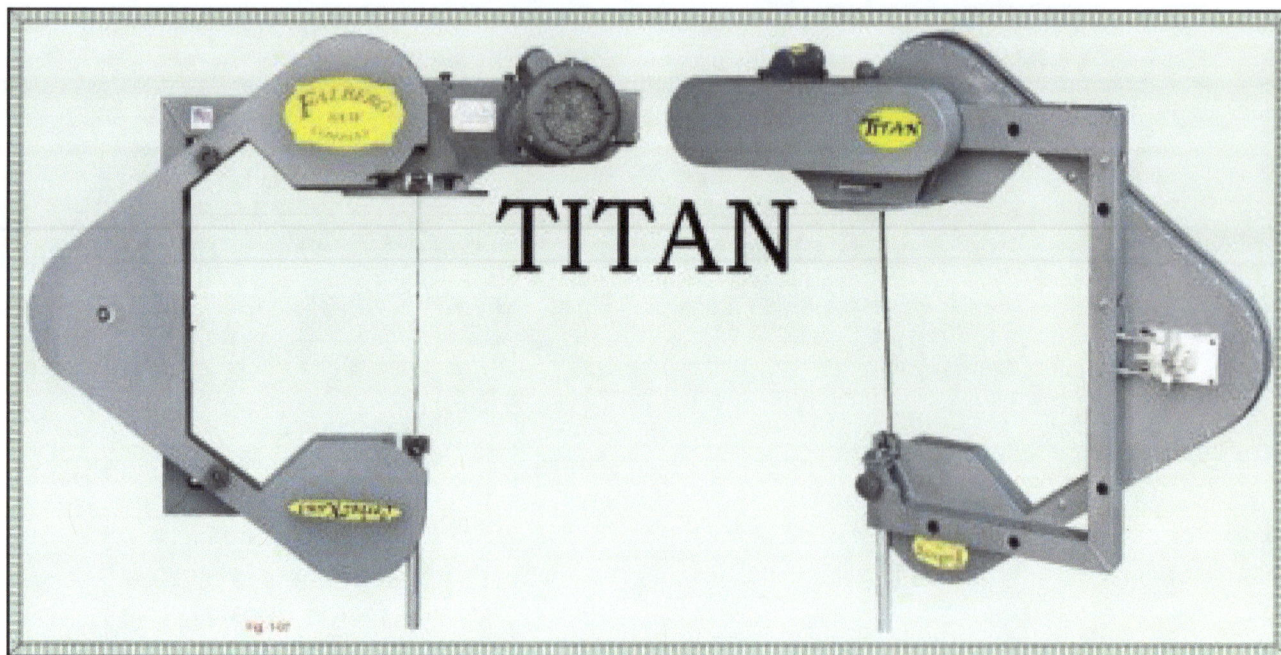

TITAN

The Corbel King is basically a Titan with a detachable table. It was originally meant to be an optional jig for cutting corbels when the workpiece was too short to practically execute with a portable band saw. Corbel Kings took on a life of their own when we started adding fences and cross-cut T's to facilitate a broader range of shop functions. A saw designed to make giant corbels has no excuse for not being able to cut the small jobs as well. (It was the inspiration for our development of the SliceMiester, which is a dedicated, high-powered, and heavier, veneer saw with a detachable fence for contour work; and is still in the prototype stage.) Corbel King stands 5' 2" tall and weighs about 100 pounds. It can cut 16.5" deep and, with the modular re-saw fence, rip perfectly straight, flat veneers with a 1/2" blade. It is not a high-speed lumber mill and doesn't pretend to be; but when it comes to the mass production of corbels, and when you need a deep-cutting saw on the job-site, there is no better alternative. Customers overwhelmingly report that the saw has paid for itself so many times over that they're too embarrassed to talk about it. They are made one-at-a-time; they are labor-intensive; and cost $6500, but that's a small price to pay when you consider the hours of work it saves dealing with big gnarly timbers all day, every day. Unless the world suddenly decides to build every new house out of timber; it will always be an expensive saw, because there's no *economy-of-scale* in this niche market.

To my way of thinking, when you've got a saw that costs as much as a good used pickup truck it ought to not only cut firewood, but wash the dishes and mow the lawn. While the Corbel King can't help you with house work, it should at least have a straight-cut feed rail system to turn odd-shapes into usable lumber and veneer. The channel-inside-a-channel rail shown below was my first experiment with feed systems and I still have one here that I use to rip flats out of raw logs. It is extremely accurate and easily portable but still has more friction than I'd like; research continues. There is a much better way and you'll see it on the new SliceMiester when I release it. The three-point end clamps also work but remain a set-up problem and they, too, will see improvement in future models. All-in-all, it's a dynamite way to start from scratch with found wood and make cool shapes. The modular fence system was our answer to the interchangeability problem one faces between setting up to cut 4" wide veneers and 16" wide veneers. How does one snug the blade guide down to a 4" veneer rip when one's fence is 16" tall? How does one rigidly support the top half and extreme ends of a

16" H x 36" L fence against aggressive feed pressure; retaining accuracy to 1/32" ; without a massive and expensive framework? How, indeed. From what I'm hearing, this must all be done with a minimum of adjustment between cuts, too. How, indeed. We'll see in Chapter Five.

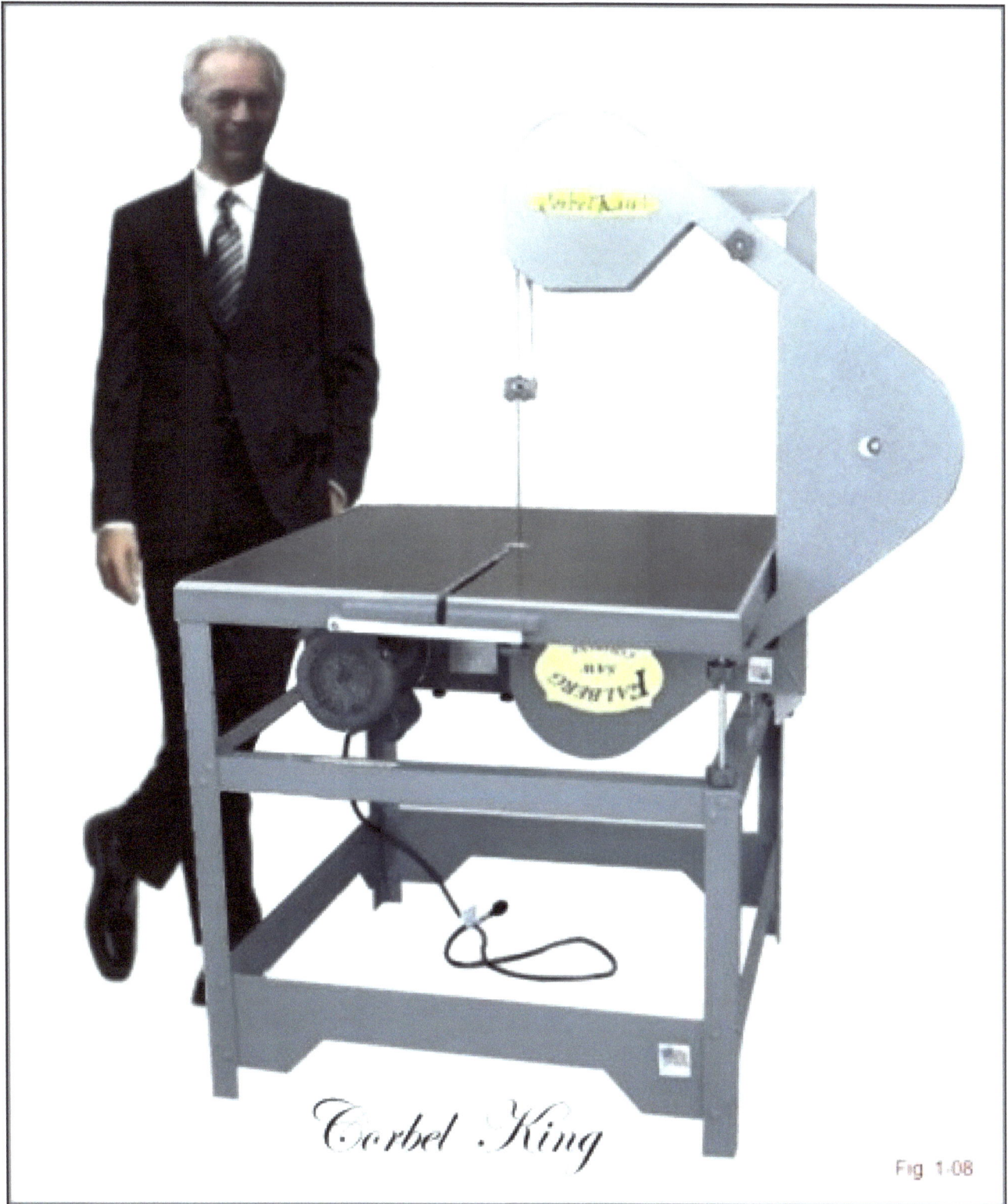

Fig 1-08

Corbel King

But first, let's talk about what all goes into the design of a band saw and how much of what we know can really be improved on. There are many compromises that must be made and I'll let you decide whose job it is to make them.

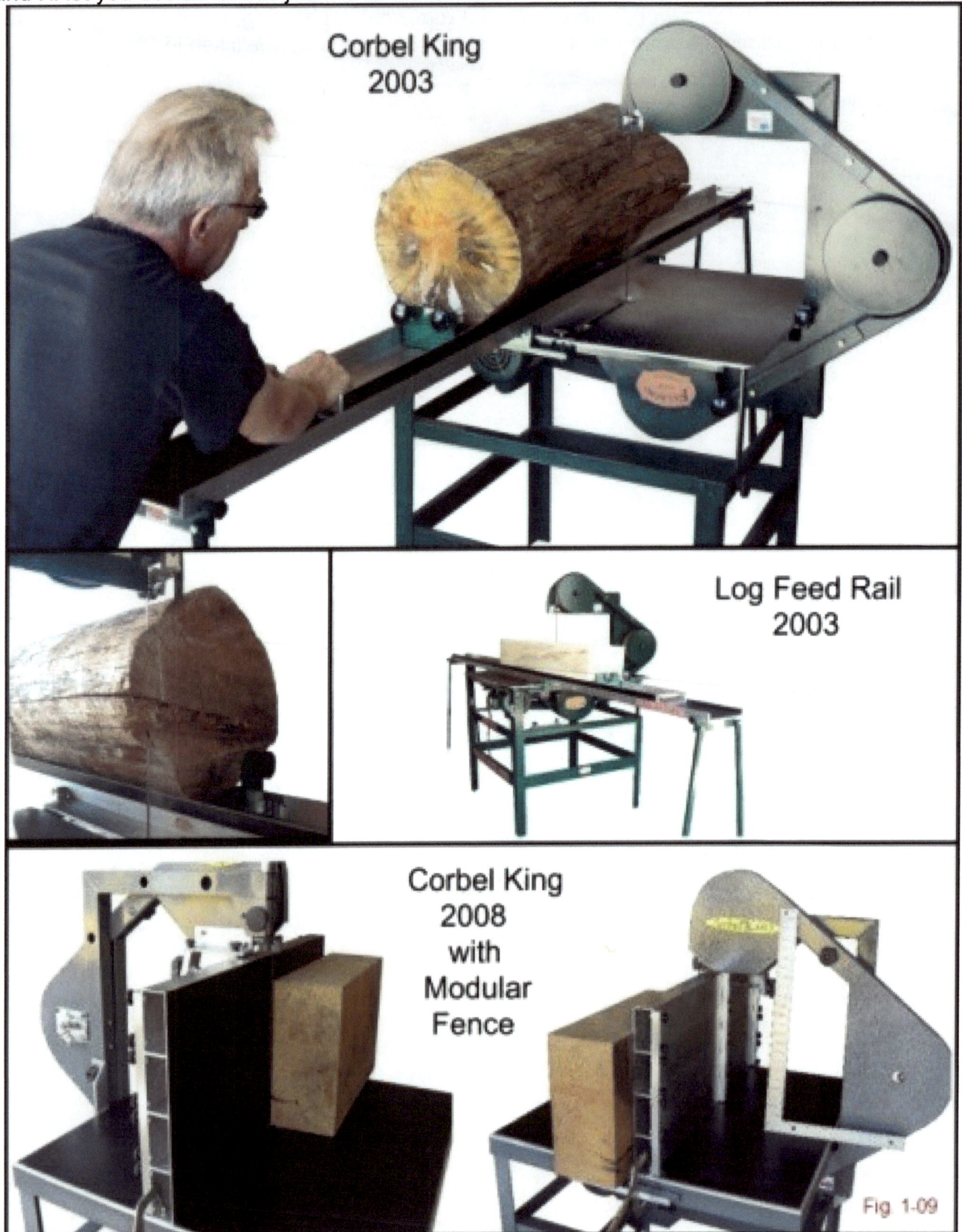

Corbel King
2003

Log Feed Rail
2003

Corbel King
2008
with
Modular
Fence

Fig. 1-09

Throughout the rest of this treatise I will try to provide a first-rate photographic record of the inception and development of band saws from a scholarly perspective but I have to warn you that getting good photos is extremely difficult on a limited budget and even more so in a warming global environment with protected species of sub-tropical mammals constantly encroaching on industrial photo-ops. I try; I really do. On the other hand, they do enhance our sense of scale and orientation; not to mention their metaphorical value for the expression of salient attributes. Pick one.

Fig. 1-10

CHAPTER TWO
Why tension is so stressful.

I was doomed from the start; obviously. My initial quest to build a big honking portable bandsaw, light enough for tight-radius timber work, left me no other choice but to use three wheels for its blade transport system. Conventional wisdom, however, is quite adamant in stating that three-wheelers are the worst form of bandsaw ever devised and that the laws of physics strictly forbid the use of such saws by all men of reason and good sense. Firstly because the massiveness of the the frame would be prohibitive; secondly because every fool knew you couldn't keep blades tracking on a three-wheeled saw; and thirdly because the wheel diameters are so small that the welds would spontaneously disintegrate from metal fatigue. I'll cover the latter two improbabilities later but for starters let's talk about how massive a frame really needs to be.

The main reason three-wheelers have such a bad reputation is that they've traditionally been designed with grossly inadequate tensioning systems. Many three-wheeled designs have been attempted in the past to produce cheap bench-top models for the hobby market but adapting the tensioning assemblies to compensate for the increased length of a free-running blade was never, apparently, consistent with their commercial objective: to make a *cheap* bench-top bandsaw. Not understanding the importance of consistent tension; they overlooked the most critical component of a three-wheeler.

Three-wheelers aren't necessarily finicky and can, in fact, be very forgiving of rough treatment. My problem at the start was that I didn't understand why. I confess to pure, dumb, beginner's luck with my first tensioner design but a hundred variations later I'm beginning to understand, and it continues to evolve.

The first open-throated frames consisted of a relatively flimsy 1/4" aluminum plate reinforced with a home-made 1/8" x 1" x 2" x 1" aluminum channel which I hand-cut from 2" square tubing. It was screwed to the frame around the blade's perimeter and served as both a blade guard and frame brace as seen below. As you can guess, they buckled in varying degrees under relatively low tension. While this type of frame flexed enough to serve as its own tension spring it was impossible to keep it tracking when the blade deflected. Not having any background in industrial engineering I had no way of quantifying the amount of rigidity the frame had, or needed to have, to properly tension a blade of given width or thickness. Further, and more to the point, I had no way of quantifying the amount of force being exerted on the frames by the tensioning springs.

I hear endless arguments over the relative accuracy of various stress gauges that break down into absurd logic on one hand and arcane science on the other. It's all good, I guess, but I come away still not understanding how it relates to cutting wood. Simpleton that I am, I just want to know how to stretch the blade *appropriately*. The factory-suggested tension values normally stamped on the saws' frames, indexed by blade-width, is a start and provides at least some guidance to the operator in search for the *right* setting. Such scales do not ascribe, however, any absolute values to tension that can be used to compare spring tension on one machine to another in apples-to-apples terms.

There is much confusion on the subject of spring tensioning and I think it derives from the improper use of the words "tension" and "stress". According to the following excerpts from Wikipedia:

"In physics; tension is the magnitude of the pulling force exerted by a string, cable, chain, or similar object on another object. It is the opposite of compression. As tension is the magnitude of a force it is measured in newtons (or sometimes pounds-force) and is always measured parallel to the string on which it applies."

"Compressive stress is the stress that, when applied, acts towards the center of that material. When a material is subjected to compressive stress, then this material is under compression. Usually, compressive stress applied to bars, columns etc. leads to shortening."

"In continuum mechanics, stress is a measure of the average force per unit area of a surface within a deformable body on which internal forces act. In other words, it is a measure of the intensity of the internal forces acting within a deformable body across imaginary surfaces."

and

".....The unit for stress is the same as that of pressure, which is also a measure of force per unit area. In imperial units stress is expressed in pounds-force per square inch (psi) or kilopounds-force per square inch (ksi)."

It's like a foreign language, huh? My interpretation of it is that one adjusts a band saw's spring compression (in pounds) to achieve the desired tension (in pounds) on the blade. Stress (in PSI) is the effect tension (in pounds) has on a blade, and varies with area of the blade's cross-sectional area (in square inches). So tension is the cause and stress is the effect. The most comprehensive understanding of cutting solutions would include knowledge of both values but the average bandsawyer knows neither, relying instead on the basic flutter test.

The flutter test consists simply of tensioning a blade until it stops fluttering and giving the tension adjustment screw an extra turn (for good measure). It works for me most of the time but when it comes to resawing there are few extra turns you might want to give the old tensioning screw; and once you're past the fluttering stage you have no other way to judge just how much extra tension has been applied, nor how much more may be needed.

The next, and easiest, step is to quantify the the amount of compressive force being applied to the saw's spring. Since band saw manufacturers never tell you the spring rate of their equipment you'll have to calibrate that for yourself. It's not hard. Spring rate is calculated as the amount of force (in pounds) it takes to compress the spring one inch. Since most band saw springs don't have an inch of free play you have to measure them in increments of 1/16", 1/8", or whatever, recording the pounds of force corresponding to each increment on an index of some sort. Transferring this index to your bandsaw will thereafter enable you to determine the amount of tension is being applied at any given setting.

To do this you'll need an automotive valve spring tester which can be bought for about $150 at any auto supply store. If you don't want to own one you could probably find one at a mechanic's garage to borrow for a few minutes. They're basically miniaturized bathroom scales and if you stand on one it'll read your weight. I bought one for myself and this is what it looks like when you measure the spring rate using a vise and ruler. Calipers are probably more accurate for this but if you measure between the vise's jaws and subtract the length of the valve spring tester's body it comes out pretty close to the same.

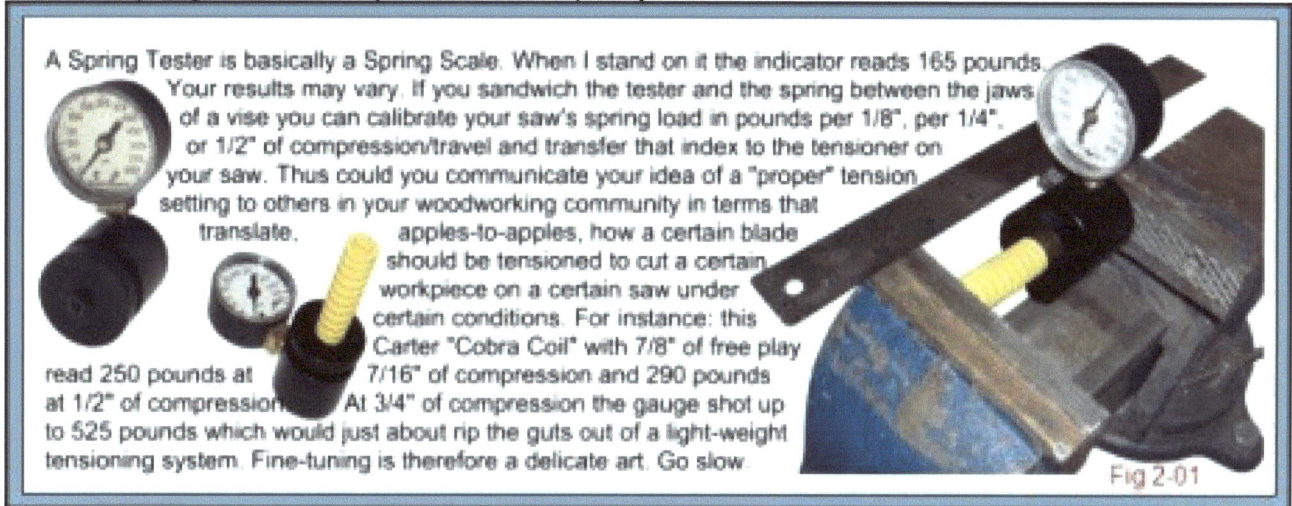

A Spring Tester is basically a Spring Scale. When I stand on it the indicator reads 165 pounds. Your results may vary. If you sandwich the tester and the spring between the jaws of a vise you can calibrate your saw's spring load in pounds per 1/8", per 1/4", or 1/2" of compression/travel and transfer that index to the tensioner on your saw. Thus could you communicate your idea of a "proper" tension setting to others in your woodworking community in terms that translate, apples-to-apples, how a certain blade should be tensioned to cut a certain workpiece on a certain saw under certain conditions. For instance: this Carter "Cobra Coil" with 7/8" of free play read 250 pounds at 7/16" of compression and 290 pounds at 1/2" of compression. At 3/4" of compression the gauge shot up to 525 pounds which would just about rip the guts out of a light-weight tensioning system. Fine-tuning is therefore a delicate art. Go slow.

Fig 2-01

Falberg saws produce considerably less tension, with the combined force of two springs maxing out at 150 pounds. With 3/4" of free play the force measurements between indexed graduations are considerably less than those of the Cobra Coil and tuning is therefore less critical. We normally run them at 3/8" of compression and the spring force at that setting is 50 pounds; right about where the Cobra spring starts compressing. Theres an astounding contrast between the level of tension my saws generate and those by saws using Cobra springs and it highlights the depth of our misunderstanding in regards to *proper* blade tension. How is it that Falberg saws will resaw 18" of timber without deflection with 50 – 100 pounds of spring tension while Cobra users are tensioning at grossly higher levels? One difference is in the width of the blades big stationary saws are designed to support. With every eighth-inch (1/8") of increased blade width you can expect to need at least another 25 pounds of spring tension to keep it from fluttering. Based solely on my casual observations and subject to more scientific testing one should expect to see a direct relationship between blade width and its minimum of required tension; that being the tension required to simply pass the flutter test. I expect that relationship would look something like this:

BLADE WIDTH	25#	50#	75#	100#	125#	150#	175#	200#	225#	250#	275#	300#
2.00"												●
1.75"											●	
1.50"										●		
1.25"									●			
1.00"								●				
.875"							●					
.750"						●						
.625"					●							
.500"				●								
.375"			●									
.250"		●										
.125"	●											

Fig. 2-02 — SPRING LOAD (lbs)

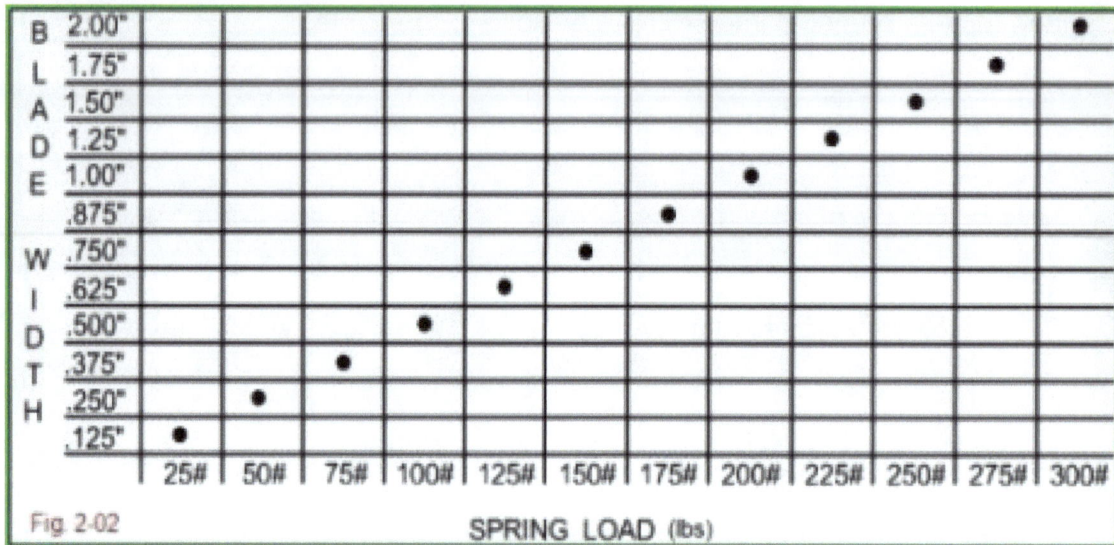

I doubt that such a linear relationship could be shown using beam strength measurements expressed in PSI. Other graphs could be constructed showing further breakdowns taking TPI values (and subsequent gullet variations) into account, for more accuracy; but I think such refinements would be minimal and the fundamental, direct, and linear relationship would stand, as shown, to be useful in the discussion of tension settings. Since spring load values are so easily arrived at by the majority of bandsawyers I had hoped it would lead to the adoption of this standard and end the "proper tensioning" debate for ever and ever. But it won't.

Some blade manufacturers recommend certain blades to be run at specified stress levels (in PSI) without regard to actual tension settings (in pounds) for their wide, narrow-kerf resaw blades, claiming it is more accurate. To do this you have to buy one of their rather expensive little stress testers. Those who don't know their saw's tension values at any given setting, which is the majority of bandsaw owners, are thus encouraged to adopt the stress test system for determining blade tension. This works also - **if** you know the factory suggested stress level: which, again, most of us don't. In either case, tension or stress, you have to somehow quantify blade tension settings if you want to establish a comparable and repeatable value for guidance and reference.

Off-the-shelf stress testers are fairly expensive, often costing more than the saws they're used on but they can be nicely home-made by skilled tradesmen such as Bob H. who created this little beauty:

Bob H. describes it thusly: "This is a ho-made tension gage. The clamps are milled from 1/2 x 1/2 steel bar. The thumbscrews are SPI Shearloc knobs on 10-24 x 1/2 socket head cap screws. The cap screws were altered with brazed-in carbide points, ground for concentricity with the threads. I built in a little magnification so that the calibration factor is 3000 psi per .001" indication. It's very repeatable and very accurate. That being said, I hardly ever use it. I have markings on my tensioning crank on the machine which relate perfectly to the gauge. Of course the relationship is for a given stress level and I chose 25K psi."

I've never understood the need for stress gauges until recently when I became involved in a WoodNet on-line forum discussion on the subject of blade tensioning. The thread was long but the crux of the matter was eloquently explained by Bob H. and two other very intelligent tradesmen who exercised tremendous patience, explaining the math and logic relating to the use of stress gauges. I'll use their words because I don't think I could explain it as succinctly.

To quote Chipper Junior: "We are just measuring strain and computing stress. Blade tension spoken in terms of stress gives a more apples-to-apples comparison as it will factor out the size of the blade. If we speak of tension in terms of force only, we will need to know the cross sectional area of the blade. A larger blade or rod area will take more force to generate an equal amount of stress. We may view a rod of any elastic material as a linear spring. The rod has length (L) and cross-sectional area (A). Its extension strain (epsilon) is linearly proportional to its tensile stress (sigma) by a constant factor, the inverse of its modulus of elasticity; pretty much E. Hooks law."

Bob H. goes on to say : "Here is the math you need:

d = actual indicator reading in inches

S = stress in psi is the stress you're shooting for (23-25,000 is good for hardback or bi-metal, but I use 15,000-18,000 for flex-back.)

L *= is the distance between your attachment points in inches.

E = modulus of elasticity in psi (I use 28,000,000 psi, some people round it to 30,000,000 psi.)

A = area in square inches (**net** width x thickness)

F = force in lbs.

d = SL/E

but S=F/A

so: d=FL/AE or FL/2 x w x t x E (Remember we're pulling on two bands--the cutting side and the return side).

" * In reference to L; the C-C distance of the attachment points is arbitrary, as long as your calibration uses the actual length. The longer you make the c-c distance, the better accuracy you'll get. I used 5.50" on my gage, but I also built in a magnification factor in the distance ratio between the indicator stem and attachment point relative to the pivot point. I was shooting for a nice round figure of 3000 psi of blade tensile stress per .001" of indicator travel. That is my calibration factor."

"Say your spring rate is 460 lbf/.250 inch or 1840 lbf/inch. Lets say you have a 3/4-3 TPI bi-metal blade .035" thick and .630" back-to-bottom of gullet, for a net area of (.022 sq. in. x 2=) .044 sq. in. To reach the manufacturers recommended 25,000 psi, you need .044 x 25,000=1100 lbs of force. You will have to compress your spring 1100/1840 lbs. or ~.6". This 1100 pounds of force will produce a strain (d) of 1100 x 5/.044 x E. Assuming your caliper is set to 5" at the start: d= ~.0045."

"PS: I like high tension! I run carbon blades at 22-25,000 and bi-metal at 25-30,000. Tri-master blades like 30,000. The higher the tension for a given blade, the more feed force it will sustain before it starts to lead. I never compensate for drift because it just doesn't happen at these tensions. This obviously only applies to straight line cutting, tension can be less when profile cutting. "

"Some people use 30,000,000 psi for the modulus of steel. I stubbornly cling to the 28,000,000 that I was taught almost 40 years ago. There have always been some discrepancies in the literature about this! Use what you feel comfy with and you'll be fine."

"Manufacturers of bandsaw blades recommend 15,000 to 30,000 psi based on a couple of things. The first are the properties of *ultimate tensile stress* and *yield stress* for the material of the backer band. Teeth don't enter into it. Secondly they (should) look at fatigue loading based on their best guess as to the minimum wheel diameter the band will be run on. The **combined** loading, pure tension, and bending (with a healthy hunk of abuse thrown in) is a more complicated calculation that we probably shouldn't get into here."

"It all boils down to pre-loading a simply supported beam in tension. The more tension load you apply to the beam, the less it will deflect when the bending load is applied. When the bending load (our feeding force) causes a negation of the pre-load on the front of the blade (and at the same time a near doubling of the tension pre-load at the rear of the blade), our blade no longer remains "in plane" and that woodworkers nemesis, *lead*, occurs. Using stress (psi) as a measure of bandsaw blade tension is just a convenient way to assign a number based on the strength of the blade. It is totally indiscriminate about machine component stress. That's where the manufacturers scale on the machine comes in. I'd like to think this scale has been designed to eliminate over-stressing of the tensioning mechanism, wheel bearings, etc., etc. So even though I like high tension, you guys with the 14 inch Deltas and clones should never exceed the 3/4 blade setting with the stock spring, or you risk damaging your machine. And **never** bottom out your spring!"

It should be noted that the participants in foregoing discussion are expert resawyermen operating above-average resaws and that Bob H. has actually designed and built several very fine professional-grade resaws. Those of you with typical14" band saws might have raised your eyebrows at the force ratings expressed in Joe Grout's stress test results and it should give you a better sense of over-all tension variance to see the spring his Minimax 16 uses. It's in another class from the Cobra spring; which itself is in a different class from the springs I use in my low-tension saws.

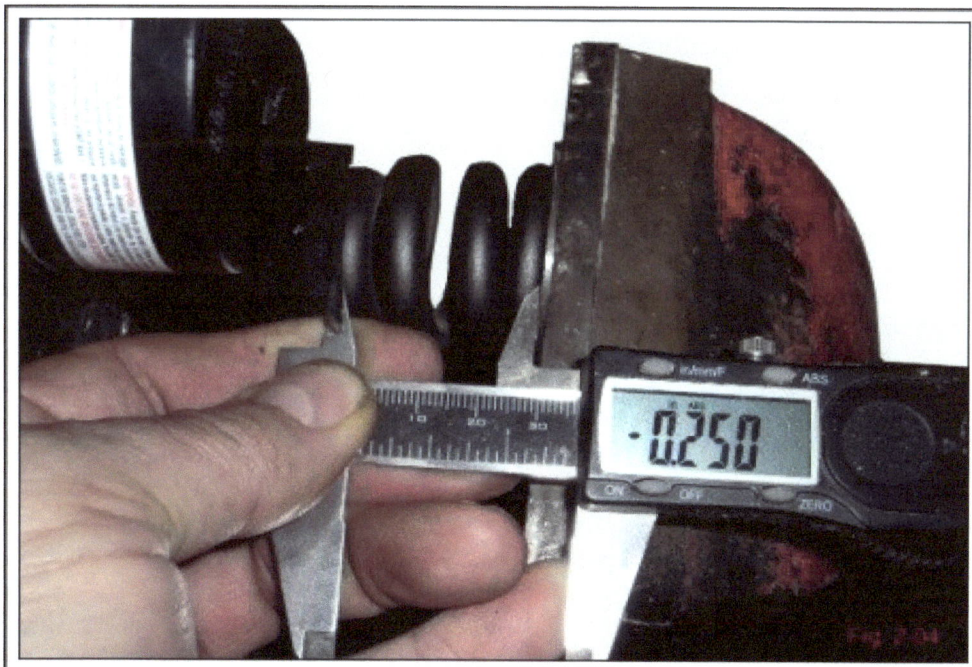

I hope you enjoyed that enlightenment as much as I did. That's what happens when real engineers get into woodworking and bring their skills with them. Special thanks gentlemen. And special thanks to Woodnet.com and all the other woodworking forums and their administrators who make such discussions possible.

But getting back to me and my weird frame designs; force per square inch, as measured in PSI, also comes into play in the tensioning spring's mounting bracket on the saw. Some of the bigger saws have the springs in-line with the wheels but consumer saws mostly have their springs mounted behind the wheels' backplate and exert their force an inch or two out-of-line behind the wheels' centers. Given the spring's location relative to the wheel's center; that force is diverted as much toward cocking both wheels inward as it is to spreading the legs of your frame. So the spring is working almost as hard to twist the tracking mechanism as it's working to spread the wheels apart. Twisting the tensioning bracket in on one end and out on the other end places the entire spring load on four small points in the slides: the top two corners will dig into the inside of the wheel housing and the bottom two corners will dig into the out-side of the slide, as shown below.

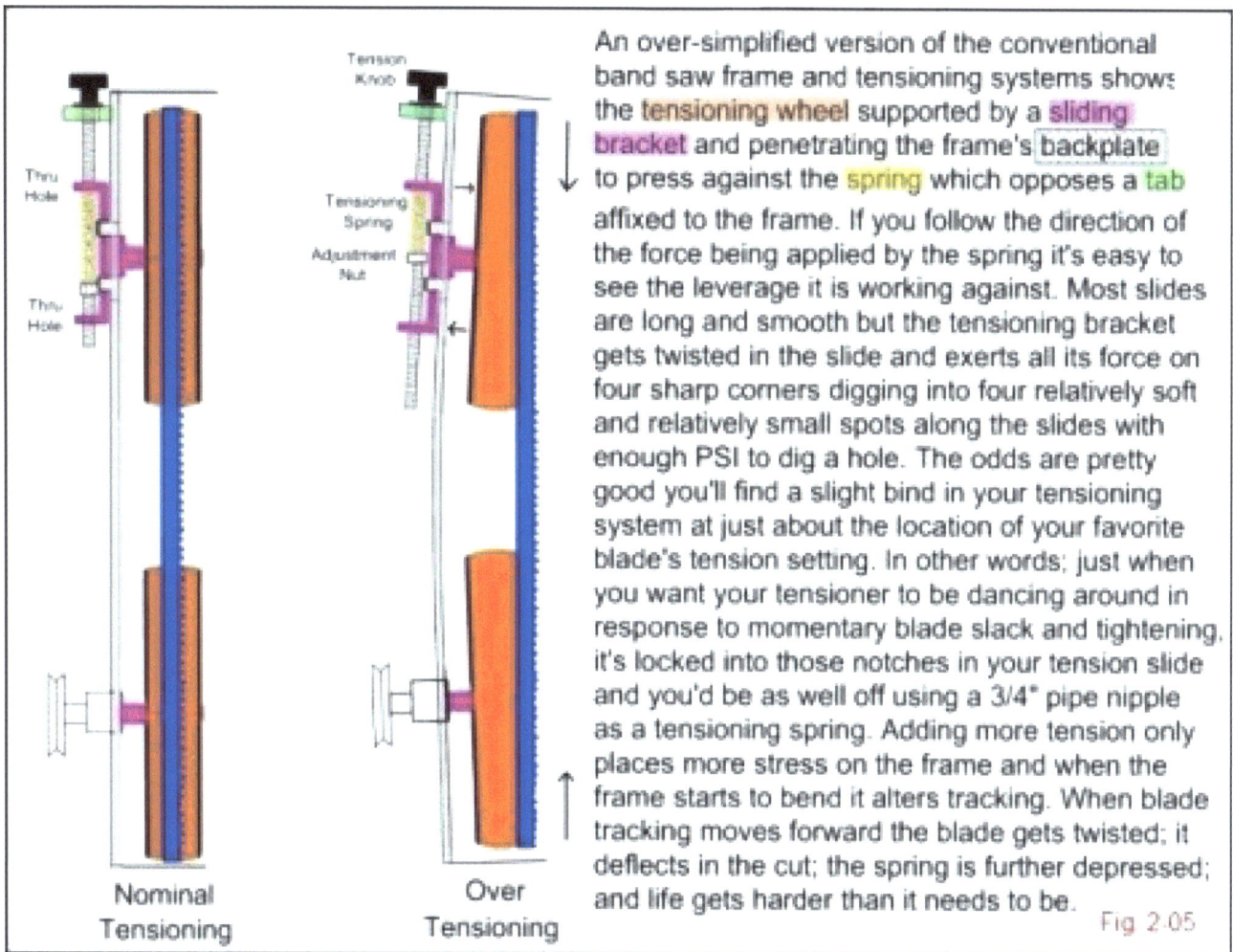

An over-simplified version of the conventional band saw frame and tensioning systems shows the tensioning wheel supported by a sliding bracket and penetrating the frame's backplate to press against the spring which opposes a tab affixed to the frame. If you follow the direction of the force being applied by the spring it's easy to see the leverage it is working against. Most slides are long and smooth but the tensioning bracket gets twisted in the slide and exerts all its force on four sharp corners digging into four relatively soft and relatively small spots along the slides with enough PSI to dig a hole. The odds are pretty good you'll find a slight bind in your tensioning system at just about the location of your favorite blade's tension setting. In other words; just when you want your tensioner to be dancing around in response to momentary blade slack and tightening, it's locked into those notches in your tension slide and you'd be as well off using a 3/4" pipe nipple as a tensioning spring. Adding more tension only places more stress on the frame and when the frame starts to bend it alters tracking. When blade tracking moves forward the blade gets twisted; it deflects in the cut; the spring is further depressed; and life gets harder than it needs to be.

Fig 2.05

Nominal Tensioning

Over Tensioning

When the tensioner's slide assembly can't dig any deeper into the frame's slide surface, it can only bend the frame itself, using all the leverage provided by its position two or three inches behind the wheels' outer bearings. Can you see how the tensioning assembly might just feel like a wedge, under the circumstances? Grease it all you want but with that much leverage digging into relatively soft metal at four opposite corners of the tensioning assembly, PSI at those points is going to be enormous and will most likely dig themselves a small notch, if not a hole. In many saws the tensioning bracket is not equal to the task and its casting will warp to cause wheel mis-alignment or break altogether. And the shorter the slider, the higher the PSI because of the leverage involved. Those little arrows behind the top wheel show the direction your spring's force is really being directed; not the same direction as the slide, is it?

Little *catches* in the tensioning slide don't sound like much of a concern, does it? But it is. For one thing; when the tensioner *sticks* the spring is no longer pushing against the wheels, it's pushing against the frame. As such, it's pushing the frame against itself. You could give the tension screw a couple extra turns but it won't increase tension on the blade, it's just digging deeper into the slides. So while you think you're applying more tension, you're really immobilizing the spring and defeating the entire purpose of having a spring. I suspect there are numerous instances of bandsawyers deriving their entire spring action more from frame flex than spring response. They will work that way and for a time my own saws were built without springs; using the frame itself to give and take up slack on the blade.

You're only going to get as much spring tension as the spring can spring. When you're *pegging* the spring at full compression, as many bandsawyers do, you're effectively bracing the tension wheel right down to the frame and the only spring action remaining to take up slack and maintain constant tension is the elasticity of your tires and what little spring action the frame might yield. That is not my idea of forgiveness, nor is it effective tensioning from the standpoint of absorbing shock, nor does it provide a constant state of tension. I hate to say this, now that you've invested serious bucks in a blade tension gauge and you've come to love your Binford so; but all the tension gauges in the world won't get you constant spring tension if the spring is completely compressed. If your spring can't bounce you might as well take it out and put a pipe nipple in there. And stop stripping the tension screw. It can't compress any more. Geez.

For years now I've preached to my customers the importance of performing the *blade-pull* test regularly to make sure the tensioning system was sliding freely. The *blade-pull* consists simply of pulling the blade sideways, like a bow, to observe the corresponding movement of the tension bolts which poke their ugly little acorn heads out from the frame. If the tension slide is functioning properly the bolt-heads will extend and contract proportionally to the amount of pull applied. My blades will typically deflect in the middle of their span for 2" to 3" which corresponds to 3/4" of free-play in the springs before full compression. The harder you pull the blade toward deflection the more tension you'll get until you reach full compression; at which point you're bending the frame. I've used the dual-spring lever-action tensioning mechanism since 2003 and every customer who's ever complained about breaking blades received the same advice: do the **blade-pull** test. The tension assembly must move, in response, instantly. They usually find that a bind has developed in the slide action and some cleaning and greasing is necessary. When they do that; the problem goes away. That's a clue, huh? You can't see it or feel it, but if the tensioner can't respond to minute deflections, it's not doing its job.

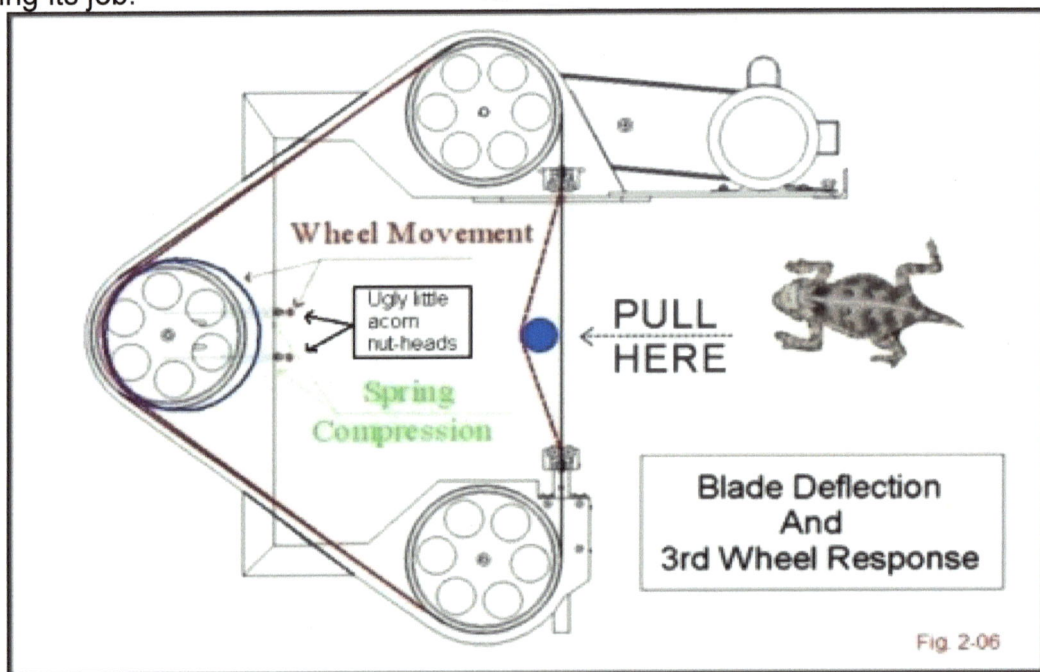

Fig. 2-06

We broke a lot of blades in the first few years and it was really aggravating to have blades fly off the wheels every time we tried to back out of a cut. What we learned, and what doomed the frame-sprung tensioning idea, was that the least amount of blade deflection affected tracking sufficiently to throw the blade off the wheels. We learned that the frame didn't necessarily have to be absolutely rigid but there was a minimum tolerance for frame flex that we couldn't go beyond. We know now that the frame must be proportionate to the width of the blade and needs only enough rigidity to compress the springs appropriate to that blade width; and that's not necessarily a four hundred pound casting. I got the weight on saw #6 down to 45 pounds, but it was too weak to run blades any wider than 3/8". My customers were demanding faster feed rates requiring the use of, at minimum, 1/2" wide blades. The process of refining each component to perform its intended function at a minimum of weight is still the name of the game here in the portable band saw business, but this is where we started:

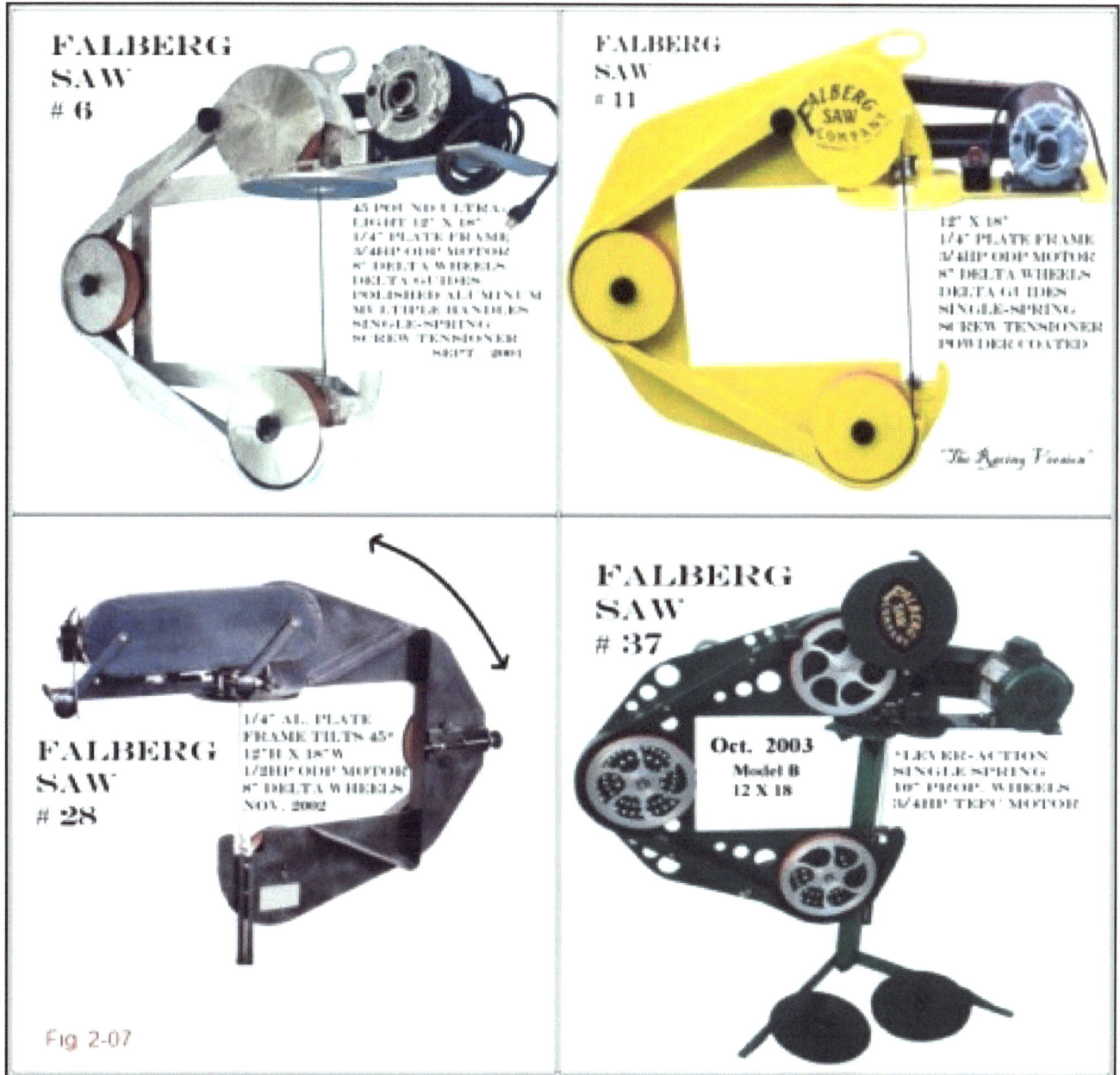

FALBERG SAW #6

45 POUND ULTRA-LIGHT 12" X 18"
1/4" PLATE FRAME
3/4HP ODP MOTOR
8" DELTA WHEELS
DELTA GUIDES
POLISHED ALUMINUM
MULTIPLE HANDLES
SINGLE-SPRING
SCREW TENSIONER
SEPT. 2001

FALBERG SAW #11

12" X 18"
1/4" PLATE FRAME
3/4HP ODP MOTOR
8" DELTA WHEELS
DELTA GUIDES
SINGLE-SPRING
SCREW TENSIONER
POWDER COATED

"The Spring Version"

FALBERG SAW #28

1/4" AL. PLATE
FRAME TILTS 45°
12"H X 18"W
1/2HP ODP MOTOR
8" DELTA WHEELS
NOV. 2002

FALBERG SAW #37

Oct. 2003
Model B
12 X 18

"LEVER-ACTION"
SINGLE-SPRING
10" PROP. WHEELS
3/4HP TEFC MOTOR

Fig 2-07

Blade deflection at the level I'm talking about now can't be seen and isn't even measurable unless you have the laboratory assets of NASA, but you have to know that your wheels aren't really *perfectly* round, your tires are most certainly not consistent in thickness, your bearings have some degree of run-out, and your blade is far from perfectly straight. Intuition tells you that there's going to be some high-speed jerkiness involved as that blade is flying around at 76 ft/sec. Common-sense logic, therefore, indicates that a tensioning system must be literally bouncing in increments too small to measure at a rate too fast for the eye to see in its effort to maintain a constant tension. What you do see is flappage through the kerf. Eliminate flappage and you'll eliminate those ugly little band saw gouge lines that makes everybody think you have to use a fine-toothed blade to get a smooth cut. Whatever you can do to regulate blade tension at operational speeds and frequencies is going to vastly improve your ability to get straighter, smoother, faster cuts with whatever saw you're using.

My fellow inventors and aspiring entrepreneurs should be all kinds of busy making after-market band saw conversion kits for hot-rodding common brands' tensioning systems. As I've said so often, with proper tensioning, you can throw your blade guides away. I hardly ever touch mine except when bullying my way through 1/2" radius turns. The SliceMiester II will use a radical new system that can't be retro-fitted to older saws but simply reducing friction on standard configurations would yield outstanding results for the average bandsawyer. If nothing else, you should check for any notches or catching in your own saw's tension mechanism and file them smooth; you'd be surprised.

My portables had to be super-forgiving of deflection to withstand the rough treatment they get in the field; bouncing off of beams while starting cuts; rocking and rolling over knot strewn irregular top surfaces; and the incredible tenaciousness they'd have to exhibit while backing blades out of right-angle turns in 16" timbers; and because they're *portable*. I'm going to get into excessive detail here because it's essential to the complete understanding of how a bandsaw works and might solve a lot of the problems you might be having with your two-wheeler. I'll remind you again – these blades are just thin strips of steel. You're driving these thin bands with enough machinery to crush rock! Doesn't it make sense that some sensitivity would be in order? Finesse? Sophistication, even?

Let's get down and talk about things the way a skinny band of steel would see this process. Let's pretend we're about one molecule tall and we're strapped down to the weld area of a blade that's about to be turned on. This is going to get real scientific now, so clean up your reading glasses and get your slide rule out. A capacitor-start 1HP induction motor goes from a stand-still to 1725 RPM in.......................oops, we missed it. Oh well, now we're whizzing along at 1725 RPM, which is 51 MPH if you're still strapped in. That's pretty fast for a short guy. If the blade you're riding is 110" long and the wheels are all 10" in diameter, you'll see the drive wheel come around every one-eighth of a second. That's eight **blade** revolutions per second. Talk about a thrilling ride! Wow!

Only those of you who ski can really relate to the forces at play here so the rest of you will have to just imagine it all. Centrifugal force is definitely working at that speed and inertia takes on a whole new meaning. Have you ever found yourself cruising down an intermediate run at 51 MPH and, for no good reason, the ground drops out from beneath you and you're suddenly, how you say...........**ballistic**; and your only means of control now is wind resistance? Depending on the terrain and your trajectory you eventually come back down to Earth with your skis pointed in the right direction, if you're as good as I am. A-hem. Anyway, that's how it's going to feel flying off a 5" radius turn and going ballistic for 30" until the next wheel comes along. But in this case it's not gravity, inertia, terrain, and wind resistance providing the thrills; it's the 1HP (arh! arh! arh!) motor, inertia, terrain (the tire), and a chunk of wood providing the thrills. Oh, yeah; and you go around eight times a second.

You question the terrain metaphor I just used? Run your neatly groomed fingers along the shiny orange crown of a brand new polyurethane band saw tire. You can not only see the spot where the tire is welded, but you can feel it. It's like the little depression on our metaphorical ski slope. If you're riding a blade weld, however, you're going"Whee!" (briefly, of course) every time that rubber-weld comes around. The same would happen if any of your wheels were even slightly out-of-round. Eight times a second. Like the expert skier you are, you're flexing your knees and absorbing the shock each time; no problem.

But if you're like Harvey Sapien, a down-valley, knuckle-dragging Neanderthal of the heavier-is-better (arh! arh! arh!) band saw persuasion you're not enjoying this ride at all. You're over-weight, stiff-kneed, and taking a pounding every time you smack the tire and still bouncing when you go air-borne again. Your skis are flapping (like an under-tensioned band saw blade) because you're being bounced around and not absorbing the shocks fast enough to maintain constant balance.

Conventional wisdom on the subject of band saw wheels would have you believe that the inertial properties of an heavy cast iron wheel smooths the feed rate and reduces vibration. They don't. If the wheels are balanced, as most are, you won't get much vibration anyway. The excessive inertia of massive cast iron wheels acts more to hinder blade stability, by slowing the spring's reaction time, than it does to reduce the flappage factor. When you consider the importance of proportionally maintaining consistent tension along the entire length of the blade while deflectional forces and workpiece resistance exert destabilizing variances, you can understand why nanosecondal response times from the tensioning system are critical. This is why race-car suspensions strive to minimize un-sprung weight. Magnesium wheels are just one example of improving a car's road-hugging ability by speeding up the wheels' response time to momentary bouncing by minimizing inertia in the suspension. Ask any sports car afficionado.

So here we go again with the stupid skiing metaphor, OK? Let's put a big chunk of wood between the drive wheel and the tensioning wheel and turn it on. Whee! You're a molecular but sentient organism stuck to the weld of a band saw blade. That's gotta feel strange. You're flying off the top, tensioning wheel at 51 MPH ; and start experiencing a distinct slow-down as you pound your way through the workpiece. At that speed it's undetectable by the operator, but the tension spring is working harder and faster now to keep the blade stretched tight between the wood and the tensioning wheel while the drive wheel is trying to leave nothing but slack on the other side. The difference in tension between the wood being cut and the drive wheel is now relatively higher than the tension between the wood being cut and the tensioning wheel. Blade tension before the braking action and after the braking action have got to be different, right? And the harder you apply braking action (feed), the faster the spring has to work to keep the tension constant. So doesn't it follow that the tensioning mechanism has to respond proportionally and instantaneously every time the blade grabs, chatters, hits a splinter, or binds on a glob of sap? At eight times per second the tensioning assembly has to bounce minutely and in proportions too small to measure and too fast to see, but its effects are manifested in snapping blades, flappage, and rough-textured cuts. Unless you absorb the shocks that happen to a blade as it travels around the blade transport wheels, you're going to get *flappage* through your kerf. That's what causes the ugly little band saw grooves you see on your veneer.

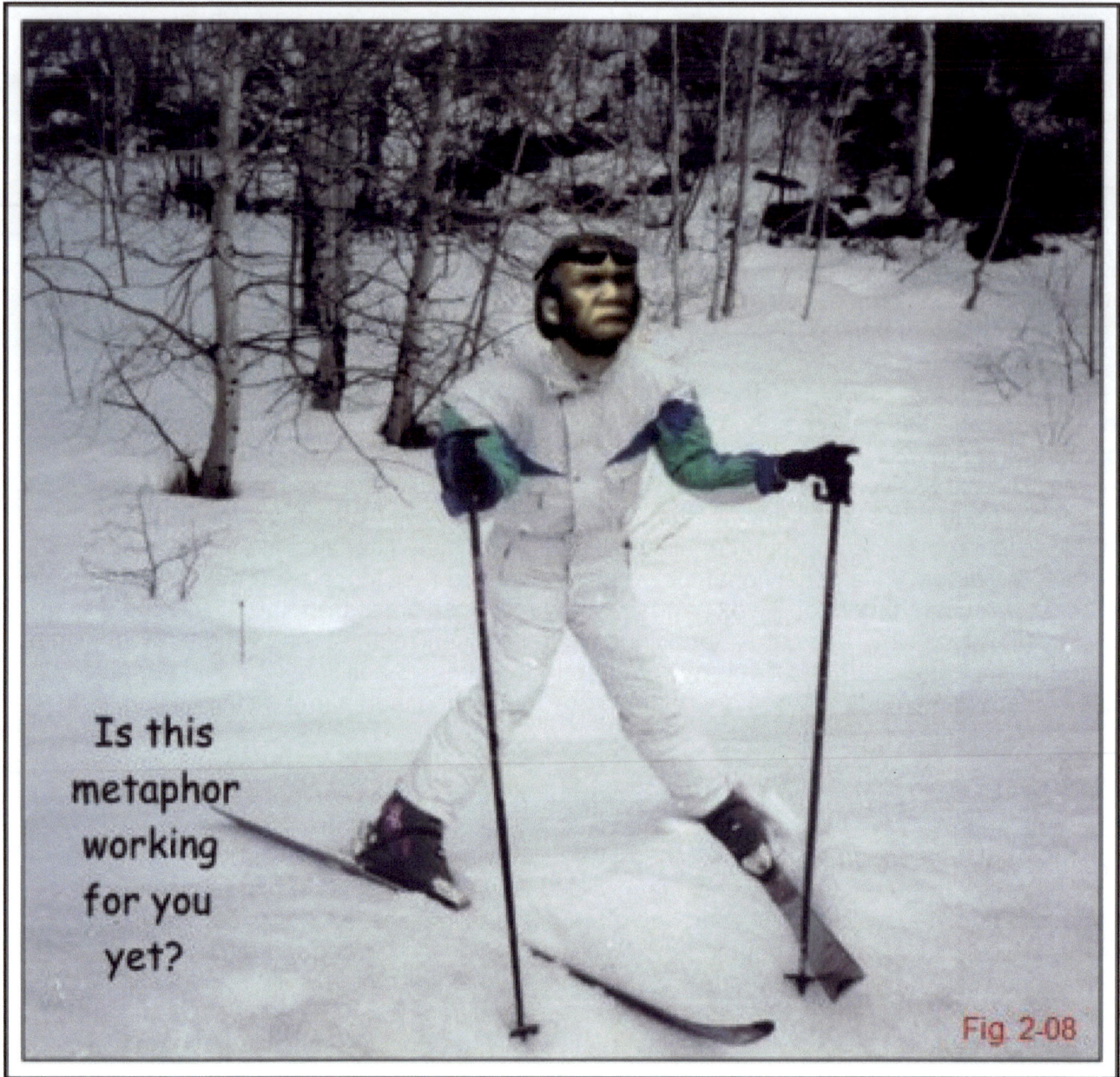

Is this
metaphor
working
for you
yet?

Fig. 2-08

Tension slide assemblies come in all shapes and sizes but they all want to bind in their slides when the tensioning assembly is *cocked*. None make any attempt to eliminate friction at the moment of deflection. Almost none provide x-y axis tracking adjustment nor provisions for the changing of springs. Nor adding springs.

Delta 20" Tensioning

Rikon gets a thumbs-up for snugging that tension spring right up under the idler shaft on this 18" system.

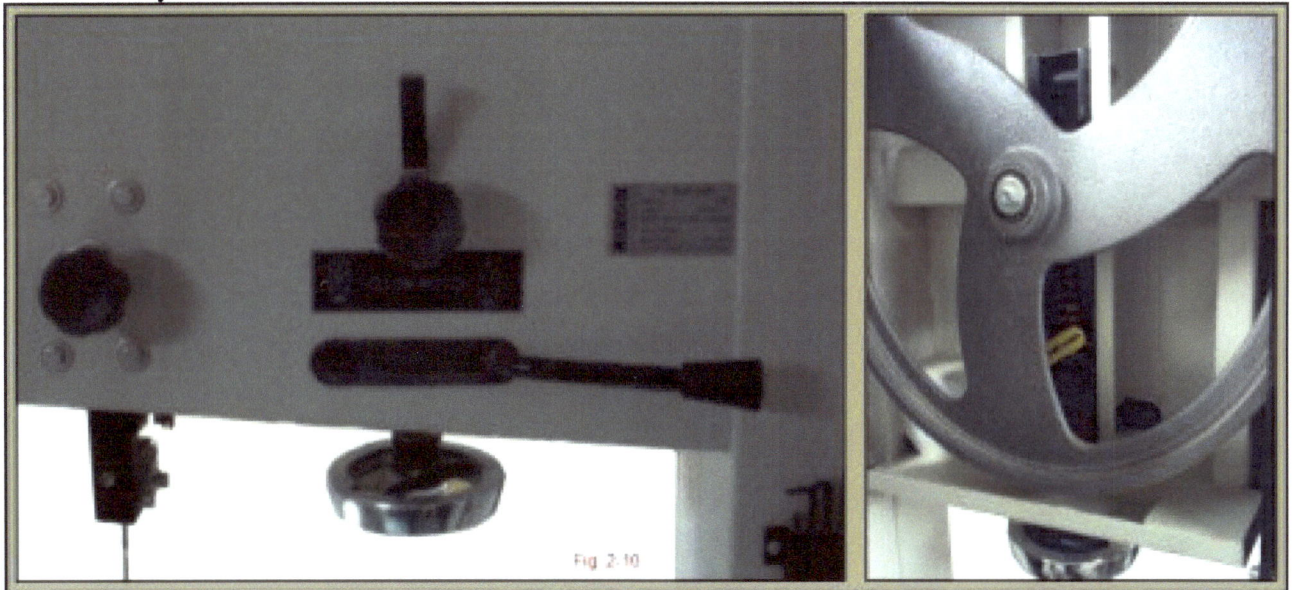

Rikon Tensioning

A variant of the basic Delta-style tension slide with a single thumb-screw controlling top-to-bottom tilt for tracking adjustment. If you have one such, be sure to keep it greased and don't over-tension; it's a vulnerability they all share. The unseen wheel-support plate is also prone to warping and cracking when over-tensioned and might incentivise replacement with a home-made improvement.

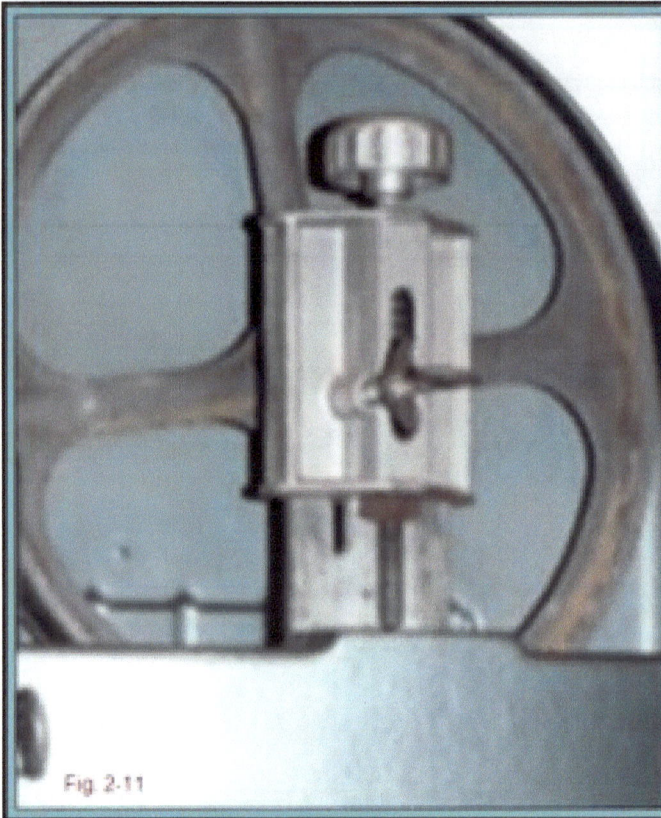

Fig. 2-11

Fig 2-12

As you can see, tensioning systems run a gamut from way simple to way complex, but they all address the matters of leverage and friction. We've come a long way since the days of wooden wheels but I think we still have a long way to go. Don't look to bandsaw manufacturers for cutting edge innovation in this field; it won't happen. But with a little design-gineering in your own garage you could not only improve your current saw but set a new standard for future saws.

Cresent

Bower

Blade tension is a relative term. You won't need near as much tension across a 3" span as you will across an 18" span to achieve the necessary beam strength for a given blade width. The longer a blade's free-span, the easier it is for a workpiece to deflect it. You can bully a blade through 3" of hardwood by choking up on the blade guides but you have to finesse a blade through 18" of hardwood with sharp teeth, adequate set angle, and efficient chip removal. Blade pitch and blade speed will play the biggest parts in determining the feed rate but set angle will determine how flat and straight the cut will be. Blade tension therefore relates more to depth of cut than it does to the width of a blade. They both relate: but to varying degrees.

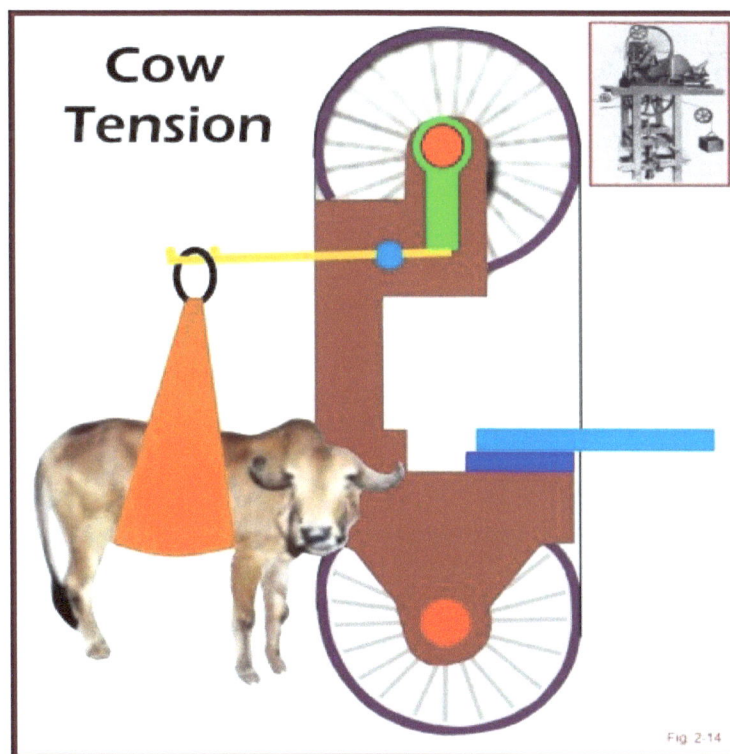

Cow Tension

Barnyard mechanics every day incorporate old ideas into new technology using the materials at hand with astounding results. As always: beauty is a question of taste, but function derives from getting it done by whatever means are available. There are a lot more means available to the average woodworker than they're probably aware of and I'll get into that in Chapter Seven.

Fig 2-15

If you haven't already found Gary Katz's "Road Show" website you should check it out. There's an episode featuring the Hull-Oakes sawmill where they use blades like this. I couldn't begin to guess what kind of tension this monster requires.

I can guess how much tension this labor of love is designed to support and it's not much. Looking closely at the tensioning system I can see why it probably cut wood as well as any other light-duty saw: its "spring" action is supplied by the frame itself. The idler wheel is simply adjusted for the length of the blade. It probably didn't turn very fast but I bet the builder/owner did some very fine work with it. I bet the wheels were taken from a older, and cracked, cast iron frame. This one won't crack; so which is better. Even when you don't know the whole story it's fun to speculate on home-made saws.

Fig. 2-16

At present, the only semi-standardized device bandsawyers have to work with are a small selection of stress gauges and, although many argue their accuracy, relevance, or validity, they offer at least a repeatable value by which to measure and compare tension settings on a band saw. It's the only value we currently have to answer the question "How much tension do I need to ...." that someone with a different brand of band saw can relate to. The problem with all of this is that the answer is expressed in PSI and applies specifically to the blade being used and doesn't necessarily reflect the amount of tension being applied to the frame. Since the remainder of this book deals with band saw design more than it does with cutting solutions the term "tension" will implicitly apply to saw frames and thus be expressed in pounds of force.

CHAPTER THREE
Re-Inventing *The Wheel!*

Everyone has a theory about wheel crowning and, by golly, so do I. You have to wonder why blades track the way they do and why some band saws have flat rims while others are crowned. The geometry of blade transport wheels is arguably the most critical part of band saw design so a complete understanding of how they work could save you hundreds of dollars and hours of frustration dealing with tracking and wheel alignment issues. And obsessive tracking adjustments are not the answer.

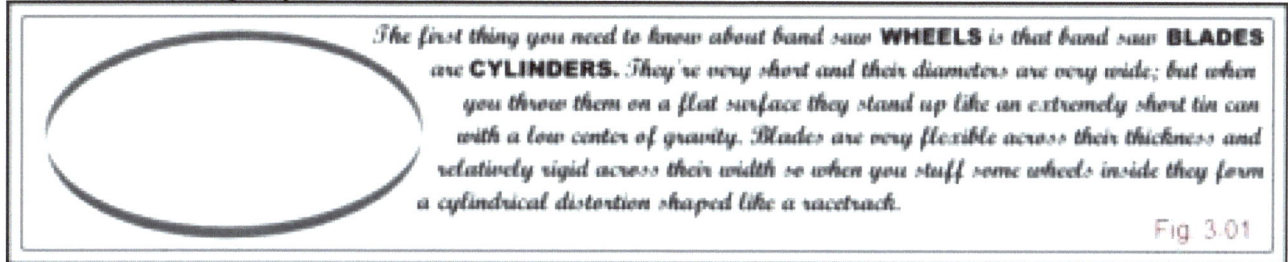

The first thing you need to know about band saw **WHEELS** *is that band saw* **BLADES** *are* **CYLINDERS.** *They're very short and their diameters are very wide; but when you throw them on a flat surface they stand up like an extremely short tin can with a low center of gravity. Blades are very flexible across their thickness and relatively rigid across their width so when you stuff some wheels inside they form a cylindrical distortion shaped like a racetrack.*

Fig. 3-01

As cylinders, blades have certain physical properties; and being of a springy steel composition they have other properties. While they're very flexible across their thickness (~.020" - .040"), they're relatively stiff across their width (~ 1/8" - 1.5"). That's intuitive and easily understood. It's in the interaction between a band saw's blade and its wheels that the relationship gets complicated. There are several forces at work when a blade tracks around multiple wheels at high speed and it's all about geometry. Of which I know know little-to-nothing in engineering terms. It doesn't matter. I'm going to try and explain it in woodworking terms anyway.

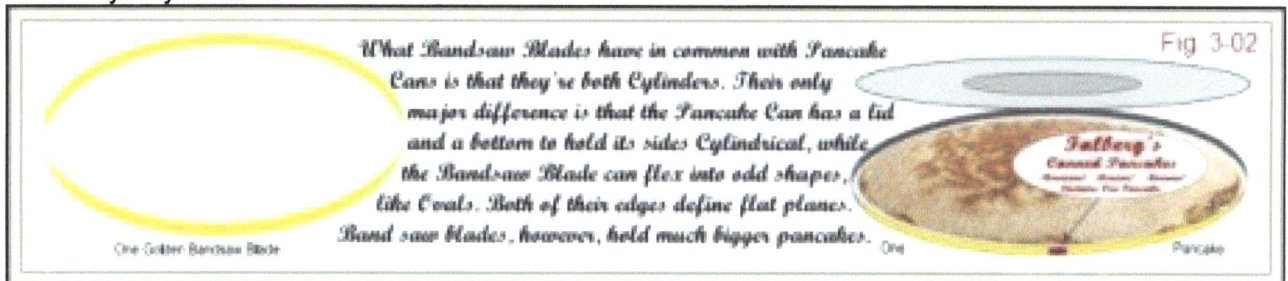

Fig 3-02

What Bandsaw Blades have in common with Pancake Cans is that they're both Cylinders. Their only major difference is that the Pancake Can has a lid and a bottom to hold its sides Cylindrical, while the Bandsaw Blade can flex into odd shapes, like Ovals. Both of their edges define flat planes. Band saw blades, however, hold much bigger pancakes.

One Gokher Bandsaw Blade

Talberg's Canned Pancakes

One

Pancake

Let's start with the concept of wheel crowning. If you were to remove the axle shafts from a traditional blade transport system and tightly wrap a rotating blade around two or more spheres that were completely free to rotate in whatever direction force was applied you might be surprised to learn that the blade will not **seek out the high spot**; nor **run uphill**; nor find any **center line** of anything. It would fall off the first time anything even nudged the blade. With the illustration below I'm asking you to imagine two spheres with mutually repulsive ball bearings at their centers trying desperately to get away from each other, but can't because they're trapped inside a cylinder (the blade).

So, while the idea of using non-axial spheres to stretch a bandsaw blade into an oval might be desirable in terms of being round, supporting the blade, protecting the blade's set, and having no axes to align; they fail miserably to restrain the blade from rolling off whenever thrust is applied. If such a spherical tracking system could exist in reality we wouldn't want it anyway and we'd have to invent a central axle to prevent the blade from rolling out of its desired plane every time deflectionary force was applied. Cutting with a band saw is dependent on the blade maintaining its axial plane in spite of deflectionary force. This is where blade tracking gets complicated, because a system with two wheels and a blade has three orbits to consider: each wheel has its own axial plane while the blade defines another, third, plane. There is a lot of latitude in how parallel these axial planes would have to be in a purely spherical system; less so in a crowned wheeled system. But not *that* much.

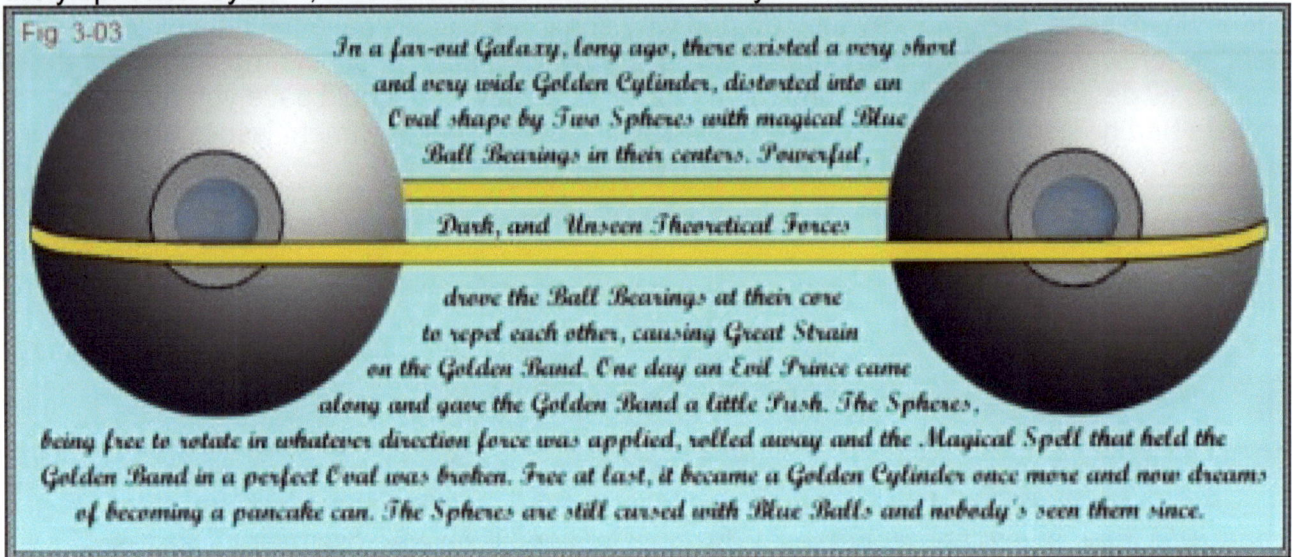

Fig 3-03

In a far-out Galaxy, long ago, there existed a very short and very wide Golden Cylinder, distorted into an Oval shape by Two Spheres with magical Blue Ball Bearings in their centers. Powerful,

Dark, and Unseen Theoretical Forces

drove the Ball Bearings at their core to repel each other, causing Great Strain on the Golden Band. One day an Evil Prince came along and gave the Golden Band a little Push. The Spheres, being free to rotate in whatever direction force was applied, rolled away and the Magical Spell that held the Golden Band in a perfect Oval was broken. Free at last, it became a Golden Cylinder once more and now dreams of becoming a pancake can. The Spheres are still cursed with Blue Balls and nobody's seen them since.

If the *spheres* above had different axial planes, as shown below, the **blade's plane** would find its mid-point between them because: as hard as one sphere would try to push the blade **one way** during its transit, the other sphere is trying to push it **the other way**. For example:

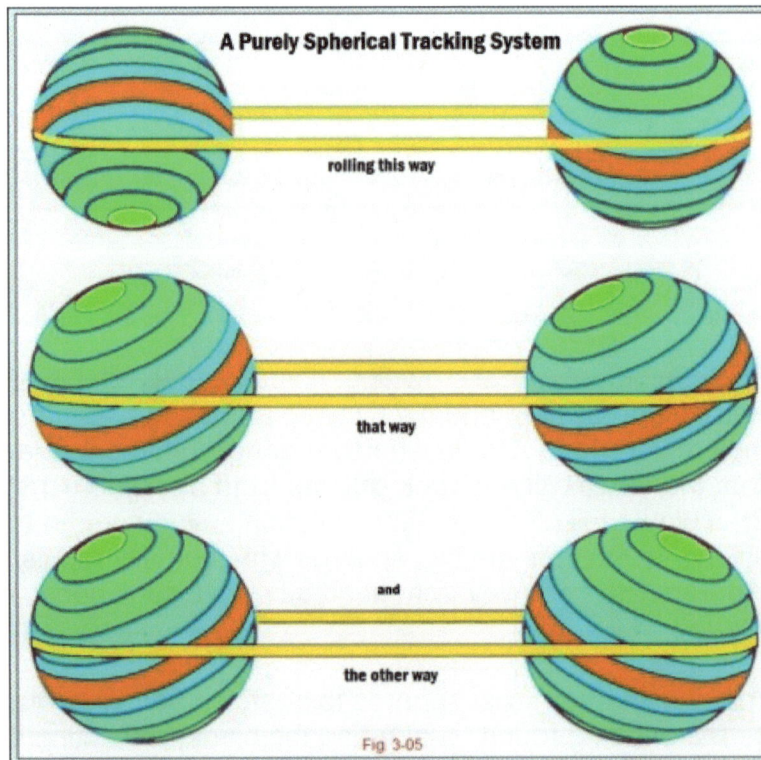

A Purely Spherical Tracking System

rolling this way

that way

and

the other way

Fig 3-05

The tracking scenarios above are quite workable and if it weren't for the obvious design problems associated with mounting a spherical system it would be extremely tenacious in operation. Based on the above you could ask why band saw wheel rims aren't wider than they are. The answer would be that so much available tracking range would exceed the limits of an acceptable thru-hole in your table and the blade guides would have to be monstrous. There are other issues dealing with the leverage factor of a longer wheel shaft, but the point is that spherically crowned wheels aren't really very sensitive to wheel alignment and, aside from a little wear and tear on the tires, a band saw can run with very little co-planar alignment if the crown is spherical.

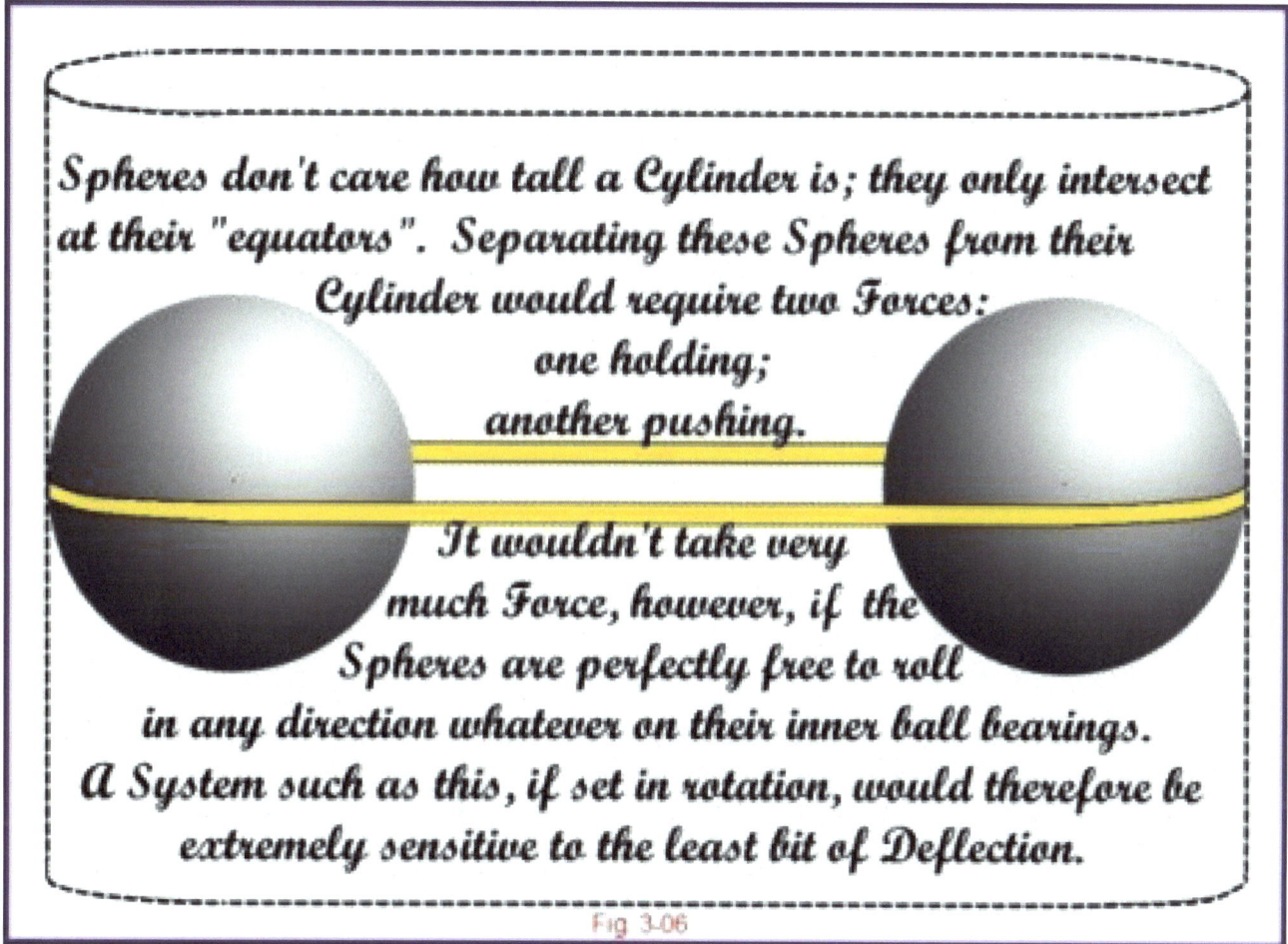

Spheres don't care how tall a Cylinder is; they only intersect at their "equators". Separating these Spheres from their Cylinder would require two Forces:
one holding;
another pushing.

It wouldn't take very much Force, however, if the Spheres are perfectly free to roll in any direction whatever on their inner ball bearings. A System such as this, if set in rotation, would therefore be extremely sensitive to the least bit of Deflection.

Fig 3-06

Pulling two rubber-coated spheres, tightly squeezed into a cylinder (albeit distorted into an oval shape at the moment), would act like a bottle-stopper in a bottle-neck. It would take a lot of force to pull it out. Spin the bottle and it would be much easier, right? That's because when you apply rotation to the bottle you're introducing another factor and I don't have a name for it but in skiing we call it "carving".

Fig 3-07

Carving, in ski terminology, is accomplished by leaning to one side and digging a hard, sharp edge into the snow. The same applies to ice skating and snow boarding. Being flexible across their thickness and relatively rigid across their width, like a bandsaw blade, they turn tenaciously; and the harder you lean into the turn, the tighter they turn. Bandsaw blades can't *lean* in any direction: they're cylinders. But wheels, if they're not crowned in proportion to their diameter, can "catch an edge" relative to the blade, and run off-track in a heartbeat; like driving screws with a hand drill.

It is this *edging* effect that makes it necessary to adjust tracking on a belt sander while it's running. As you know, this can be a very sensitive adjustment and requires the precise alignment of flat wheels in an even taller cylinder. Tracking wide bandsaw blades on a flat rim requires the same amount of precision but it's do-able on that scale for the same reason belt sanders work. Tracking a 1/2" bandsaw blade on flat wheels is possible but it would be a nightmare to keep it tracking while the blade is catching edges with every deflection during operation. It is for this reason that blades become more sensitive to proper crowning as their widths decrease. Try tracking a 3/16" blade on flat wheels. You might have to close the doors and windows to prevent drafts from blowing the blade off-track. I think the tracking problems so many woodworkers are having with their bandsaws derives from improper wheel crowning in the sense that even though their wheels are crowned; they're either too crowned or not crowned enough. Either condition can cause the blade to *edge* into a tire. *Edging* is what causes blades to appear to climb crowns to their high point. That's why you won't see band saw (blade transport) wheels shaped like V-pulleys.

Most bandsaws have an adjustment for top/bottom alignment and you can adjust the blade to track on the wheels' top and bottom centers, but I don't know of any that allow for side/side alignment (except my own; ahem). Side-to-side mis-alignment has little effect on properly crowned systems, however, and it's usually ignored. To say that a blade "wants" to find the crown, or climbs to the "highest point", or "drifts to center" is about as silly as 17th century philosophers describing gravity as an object's natural "longing" to return to "mother earth". Watch closely what happens to a blade's edge when deflectionary forces are applied.

Now, on the other hand, if you want to control where a blade is going to ride on a wheel's rim, and deflection is not an issue (as is the case with big band mills); flat rims are the way to go.

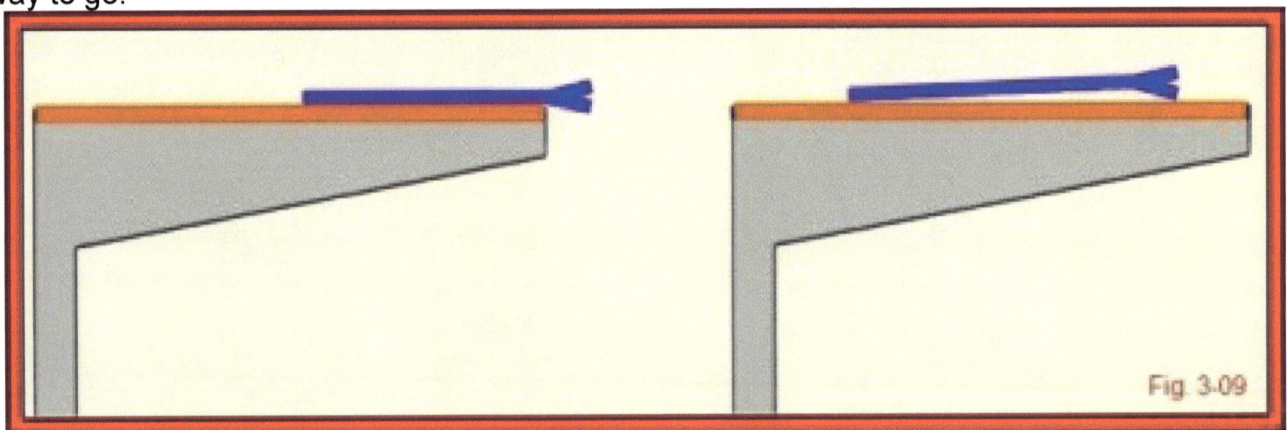

Fig 3.09

Flat rims have nothing to do with supporting the blade in any sense. Flat rims allow one to control the blade's plane by using the blade's *edging* properties to turn the blade left or right by *leaning* as a skater or skier would, to *carve* a turn. Or as you do to adjust the tracking of a belt sander. Big saw mills, using blades of 2" or more, have so much tension on them that rubber tires would be crushed; and their dimensions are so huge they can adjust co-planarity very precisely without resorting to micrometers. Big saw mills are often set up to run with the teeth over-hanging the wheels' rims. Even smaller mills are designed to run with the teeth over-hanging the rim because the tooth-set would be rolled out by the wheels' rims, if they were flat. There is no reason I can think of to use flat-rimmed wheels other than the manufacturing difficulties involved in machining wheels of that diameter. Contemplate the crown radius of a 42" wheel; you'd barely see a crown at all. If the blade's set were to exceed the crown's radius, however, you'd have to over-hang the blade; and in that case you'd have to use *edging* to control the blade's tracking. If *edging* is therefore desirable, flat wheels would be the easiest to control and adjust. You would do so with the saw turning and at the scale big saw mills operate, there's little deflection to run them off-track. The choice between crowning and not crowning gets confusing in the 1" to 2" range of blade widths and I suspect most of those with saws in that range would be better served by crowned wheels. The amount of set you usually see in blades of that range are well within the crown radius of the saws they're riding on. The idea of bandsaw blades needing "support" is highly questionable. I've never seen a cupped blade for lack of "support" as shown below (right). Sanding belts are paperbacked and need support across their entire width but band saw blades do not.

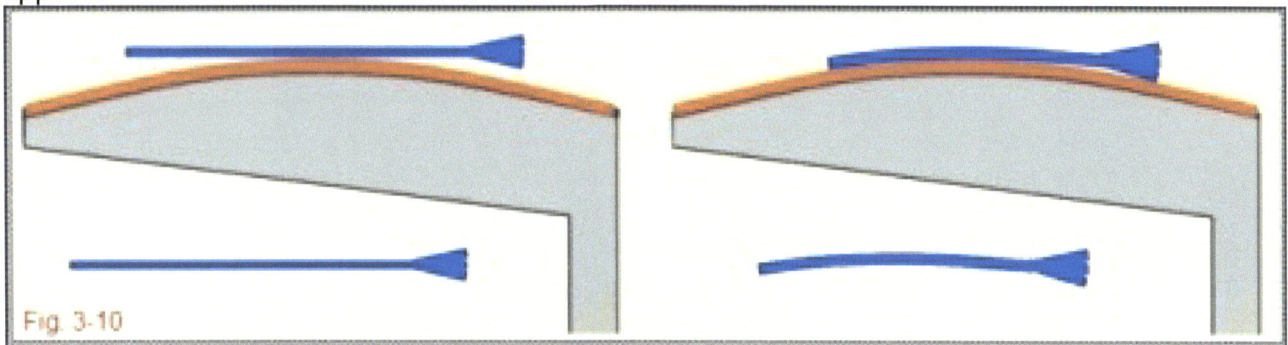

Fig. 3-10

Which of the above drawings best describes an unsupported blade?
I've never seen a cupped blade so I vote for the one on the left.

But while I've never seen a cupped blade, I have seen the set rolled out of blades by over-tensioned blades on a flat or under-crowned rim. I think this might be the cause of many complaints that a blade, all of a sudden, drifts consistently to the left or right. The wheel-side set can be rolled out by worn, grooved, or under-crowned wheels. If you doubt the ability of a bandsaw wheel to roll the set out of a coarse-toothed blade you should take a pair of needle-nosed pliers to your 2TPI blade right now and get the feel of it. I'll wait here while you do that………………………

Now, see how easy that was. I re-set bandsaw blade teeth here quite often and it doesn't take much force. The levers on my blade setter aren't as long as the handles of your needle-nosers and I do two teeth at a time. The blade steel at the base of the gullets is as soft and pliable as the steel in the body of the blade. The tips of the teeth are extremely hard and brittle, however. They break off in high-speed chips which will actually sting your skin, leave welts in your arms, and do serious eye damage amazingly fast if you apply a little pressure to them with hardened steel. Little, tiny chips; you can't even see them, but the blade won't cut anymore, and you're wondering what happened. That also happens when your blade runs off to the side of a ridged wheel and the tooth set rolls over the little ridges on either side of the tire. For that reason you should also make sure that the crowns of your tires are higher than the ridges that surround them. Tires have every right to be *proud*.

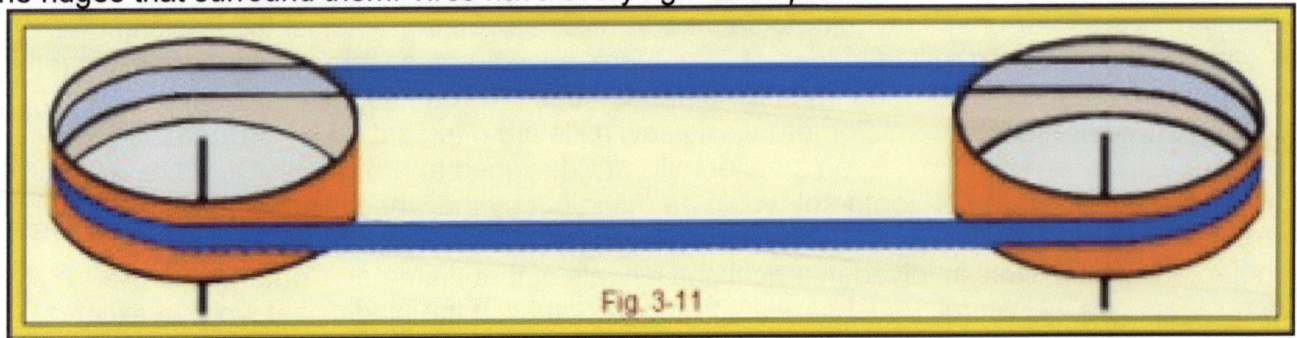

Fig. 3-11

If the wheels are in perfect planar alignment and the blade is perfectly cylindrical, it will rotate around two perfectly cylindrical (uncrowned) wheels. In proportions like those of a timber band mill it is possible to achieve such perfection. On smaller saws, where dimensions and angles are harder to see or measure, that level of perfection is very difficult to achieve. Minute variations in bearing run-out, blade distortion, and frame-flex that show as measurable and correctable phenomena in larger dimensions become impossible at the scale of 1" or less blade width and it becomes necessary to compensate for all these little variables with infinitely incremental adjustments to spring tension, blade guides, and x – y axis wheel alignment. Even if you could get the wheels into perfect co-planar alignment, the blade would still want to alter its tracking every time you applied stress because blades of smaller width will stretch and distort more than blades in the range over 1 – 2 inches and run off the wheels.

If the wheels were flat-rimmed, the blade would enter the wheels' planes *on edge* and *ski* into whatever direction the blade made contact. There's no *natural tendency* to find the crown or *high spot* of a wheel, it's the work of the devil! Band saw blades on band saw wheels act just like skis on snow, skates on ice, and bikes on sidewalks: just lean into the direction you want to go.

I've watched my blades' tracking performance on the bench pins: motor running; blade engaged. If you take a prod to the backside of the blade, in the cut path between top and bottom wheels, and pull it forward as if to escape the blade guides, the blade will track forward on those wheels and to the back side of the outside wheel. Repeating that procedure anywhere in the blade transport circuit shows that the *blade* can shift its entire axis in response to deflection, so long as it remains on the wheels.

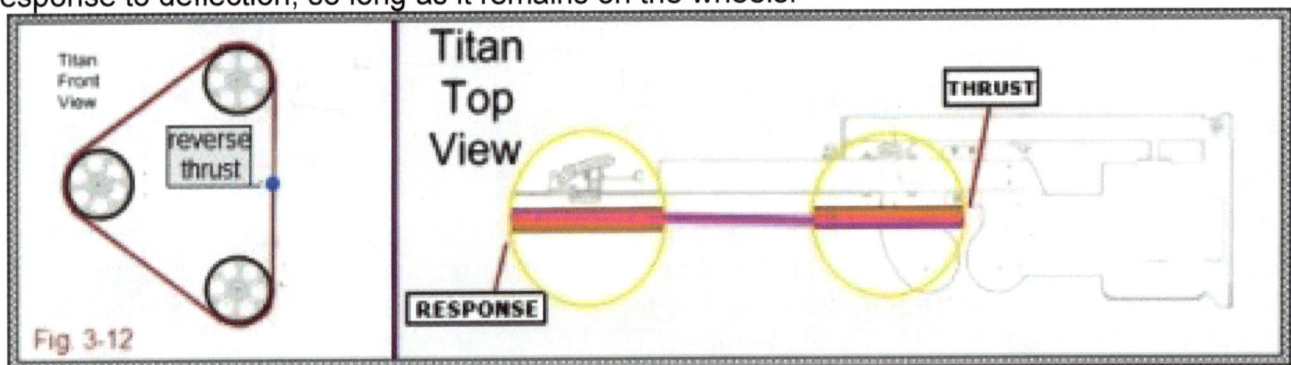

Fig. 3-12

To further illustrate the logic of spherical crowning I drew some other whacky tracking scenarios, as seen from behind the spine that look like this:

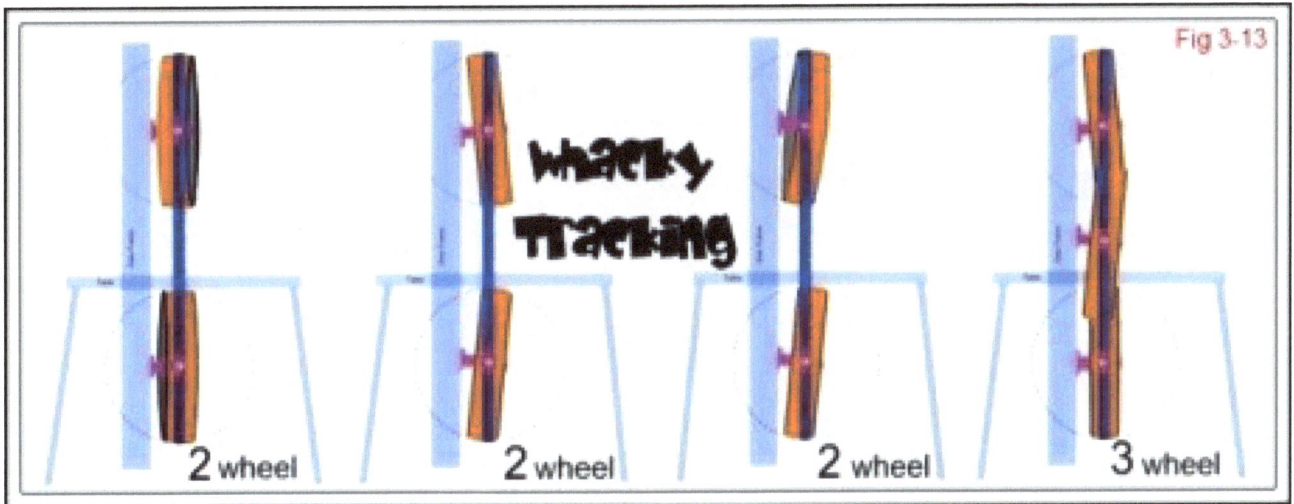

Fig 3-13

2 wheel   2 wheel   2 wheel   3 wheel

Since I design and build band saws of both the two and three wheeled persuasion I get to see some really whacky combinations of co-planar mis-alignment. In the final assembly phase, before I fine-tune the wheels' co-planarity, I turn them on to see if anything is rubbing, rattling, flapping, or smoking. My saws are designed for x – y axis tracking adjustments to every wheel and until they're adjusted, they can be way out of line. It never ceases to amaze me how bad this situation can be and still have the blade tracking; under power. Look closely at the fourth drawing above and try to visualize the blue blade entering into the plane of the middle wheel. Can you see how the blade's plane is running cross-plane to the middle wheel? You'd think it would want to follow the tire and veer off to the right, throwing the blade off track, right? No, it doesn't; because as hard as the tire of the middle wheel is pulling right as it enters that wheel's plane, it is pulling left as it exits. If the wheel-rims had been flat, the blade would *edge* immediately off the wheel. Granted, such mis-alignment is not *goodness* in terms of mechanical efficiency, but a band saw will *idle* like this all day. As you can imagine, such mis-alignment as shown above would have a narrower *margin of error* when subjected to operational deflection because they're already tracking so close to the rims' limits. Spherically crowned wheels in co-planar alignment, however, will capture the blade more tenaciously than you'd think possible.

There's nothing magical about it; if the combined average of the wheels' planes coincide with the blade's plane it will tend to stay in that alignment unless you exert sufficient force to the blade to push it outside the combined wheels' average plane. If the wheels are aligned co-planarly, the width of the wheels' plane would be one inch; the width of a tire. Rubber tires are the glue that makes sliding that cylinder off the plane of the wheels very difficult. It's like trying to pull one of those expanding-rubber bottle-stoppers out. You can do it, but it's not easy.

Preservation

Fig. 3-08

There seems to be a great deal of confusion surrounding the behavior of wide blades (>1") versus the behavior of narrower blades (<1") . To hear the experts tell it, they adhere to entirely different laws of physics. Not so. First of all; like I said, band saws don't have to have crowned wheels. They can, and do, have crowns that vary from the spherical model I use. On two-wheeled saws it's quite possible to use flat rims like these shown below and, at least theoretically, you could use them on three-wheelers as well if you had the patience to track all of them perfectly and promised to never, ever, let the blade get deflected to any degree. Flat rims, because of the foregoing caveats, are most often associated with re-sawing with wide blades. Re-sawing with wide blades almost always means big wheels, which means you're working with massive saws and lots of spring tension. I don't know of any reason you couldn't crown a big wheel to be forgiving of deflection but what would be the point? Deflection rarely occurs at that scale to be significant, proportionally, but some crowning is necessary if you intend to track narrower blades on the center of the tires of said big wheels, especially at high tension. So we can infer that the crowning of some big-wheeled saws is meant to present the option of running narrower blades than the saw was built to handle. A nice option, I would say. It makes no sense to limit a saw's adaptability to narrow blades if you can just as easily crown the wheels enough to avoid mashing the tooth-set out of narrower blades.

You can make a band saw to track wide blades on flat wheels like a belt sander does, but the tracking would be extremely sensitive and it would need to be adjustable while the power was on. You could thus control the blade's orbit and position on the wheels but it would be a four-armed nightmare to do so on a three-wheeled saw. Such set-ups would have no tolerance for deflection whatsoever. I think a lot of the cheapie three-wheelers weren't spherically crowned and exhibited the sensitive tracking characteristics of a flat wheeled saw. Flat-tracking, or insufficiently crowned wheels, would explain the tracking problems so many have experienced with three-wheeled saws. If your saw seems persnickety it might be because the crowns are too shallow. A good way to check this is to hold one wheel edge-on, the other wheel perpendicular, and visually compare the two arcs. They should be the same.

It's all inter-related with band saws and you can't make one change without changing everything else. A saw that can't be adjusted in every dimension is going to be a real headache if you're doing a wide range of operations, as most of us do. So here's another compromise to consider in designing consumer band saws: bigger wheels for thicker blades or smaller wheels for wider set? Given the significant value of a spherically crowned wheel it follows that smaller wheels will run wider blade **sets** at higher tension without *rolling* the set out of the teeth. You can do it either way but gaining any advantage always means giving up another. The only way you can win is to know what your saw can and cannot do well. Not every saw is going to work on every blade that works on certain jobs at the speed you'd prefer or the blade budget you had in mind. I'm just saying............there are dark, mysterious forces at work here and the emptor must be caveat. God knows I try to be venditory.

Blade tracking is a much bigger issue with portable bandsaws than it is with stationary bandsaws. First of all, stationary bandsaws have larger tables on which to support the workpiece so if it has an uneven bottom surface the deflection caused by rocking is minimized by spreading the angle of deflection over a wider area. The Titan's platen has only 11" of diameter and when it rides over a protruding knot as little as 1/16" high it tips the whole saw by one or two degrees. It doesn't take much tippage of the platen to create a 1" deflection at the bottom of a 16" deep cut. We've already seen what blade deflection can do to the tensioning mechanism when we did the blade pull test so imagine how compressed those springs are going to be with a sustained 1" deviation trying to work itself back into the nominal 90* kerf. Secondly; Titan's blades are subjected to a terrible beating every time you start a cut and jockey the 70 pound beast into position. It's unavoidable. There's always that instant when you ram a fast-moving blade into a monstrous timber, usually at the the most acute angle. If you've never operated a Falberg saw and seen the tenaciousness of its tracking you can't truly appreciate the importance of equating the arc of the crown to the radius of the wheel.

Coupled with the myth of small wheels' excessive weld fatigue is the myth that three-wheeled saws subject a blade to 33.3% more bends per revolution. That's true. But when the blade is 33.3% longer it's all the same. The weld still only comes around once per revolution and it still works out to the same number of bends per second. It doesn't matter how long the blade stays bent going around a wheel; once it's bent, whether it's one third or one half the way, it's bent. That's it for the duration. It gets more complicated as you work through the math so I'll leave that to you. The important thing to keep in mind when contemplating weld fatigue on two-wheeled vs three-wheeled saws is blade revolutions per second, not wheel revolutions per second. Three-wheelers have fewer blade revolutions per second because their blades are proportionately longer. They make turns around a wheel at about the same rate. If the wheels are a little smaller, the fatigue rate is only marginally increased. Balanced against the over-all benefit you get from the three-wheelers' dramatically wider throats it's a no-brainer; three-wheelers give you more bang for the buck. The blade, in a controlled, fence-guided environment, is going to wear out its teeth before the weld wears out. I've seen little evidence to support the notion that wheel diameter affects weld deterioration to the point where it should be a critical factor to band saw design. Blade deflection aside, any wheel diameter will support a 3/4" or narrower blade for a life span well within the limits of normal tooth wear.

In the beginning, before the fabulously expensive $200 Falberg 10" wheel, there was the die-cast pot-metal $17 Delta 8" wheel. There were other, narrower, wheels of course; but at 1" width, Delta had the only rim wide enough to track blades as tenaciously as I needed for my portable saws. Although widely trashed by *anyone who knew anything about band-saws*, the 8" wheels worked quite well on my earliest prototypes and I'd still be using a modified version of them today if they were still available, which they aren't. Contrary to the opinions of the world's *leading experts* (who never actually **designed** a saw, BTW), the 8" wheel is totally adequate for blades of 1/2" or less. Any breakage we encountered in those early years was totally due to the tracking difficulties inherent in our frame design.

Three-wheeled saws have a totally unfounded reputation for breaking blades and most of that myth is based on the false assumption that their smaller wheels cause undue stress at the weld joint. In view of my experience with 8" wheels during the first two years they were not a problem. The blades would get dull and we'd have to replace them after 20 - 30 feet of tough cutting anyway. A lot of that, no doubt, had more to do with the fact that I was using 3/8" low-tension blades and dulling the teeth with Doug Fir before they had sufficient time to fatigue the welds; but in my shop and out in the field, on my saws, the blades were just as likely to break anywhere as they were to break at the weld. The myth persists, nonetheless, because it's so believable in theory. The numbers: comparing a 10" diameter to a 14" diameter suggest a vast difference in the amount of bend a blade is subjected to, but when you compare the most relevant numbers, the radii; at 5" and 7" the difference is less dramatic.

The 8" Delta wheel isn't shown in the illustration below but our experience with them supports the same conclusion: wheel radius isn't near as significant a factor to weld fatigue as blade guide and workpiece deflection. So the real question is: at what point does wheel radius become critical to weld life-expectancy. It would benefit nobody to have a blade's weld last forever when the teeth are going to dull in four hours anyway. You can argue that bigger is better as far as wheels are concerned but exactly how small is too small: at 6", 8", or 14"? I've seen industrial applications of four-wheeled saws with 6" wheels in constant use with no mention of weld fatigue. Manufacturers of saws with 48" wheels can look down the chain and denigrate anything with wheels less than 48" in diameter as being intrinsically *hard on blades*. And so on down the line until you get to Falberg saws, with 10" wheels, and say "They are too hard on blade welds!" But where is that statement born out by fact? Nowhere. The weld fatigue myth as it pertains to wheel diameter has no basis in fact; but Omni-Megacorp Bandsaw Co. advertising can (and still does), attribute blade-weld failure to insufficient wheel diameter. Tch.

There were, and are, other 8" and 10" wheels but they're all too narrow for our purpose. To function in the working environment our saws are subjected to, the blade needs room to move side-to-side along the wheels' crowns. You may have seen the blade coming totally out of the blade guides in some of our videos demonstrating rafter-tail cuts. We often back out of a cut to take second, third, fourth, or fifth passes to *file* out a 90* corner. I don't know if any other saw can do that, but it still amazes me. Prior to the introduction of the dual-spring tensioning system, such maneuvers frequently pulled the blade so far off-track it escaped the wheels, often breaking the blade; but for its time the old single-spring system exhibited tenacious tracking capabilities under the stress of rough handling. I attribute that tenacity to the spherical crowning of our wheels and the mechanical efficiency of our tensioners . Tenacity, in this case, refers to the range of blade deflection within which the blade will maintain relatively constant tension and continue to track on the wheel's crowns under adverse conditions. Our saws are frequently used to cut right-angle turns in tight quarters so tenacity is a prerequisite.

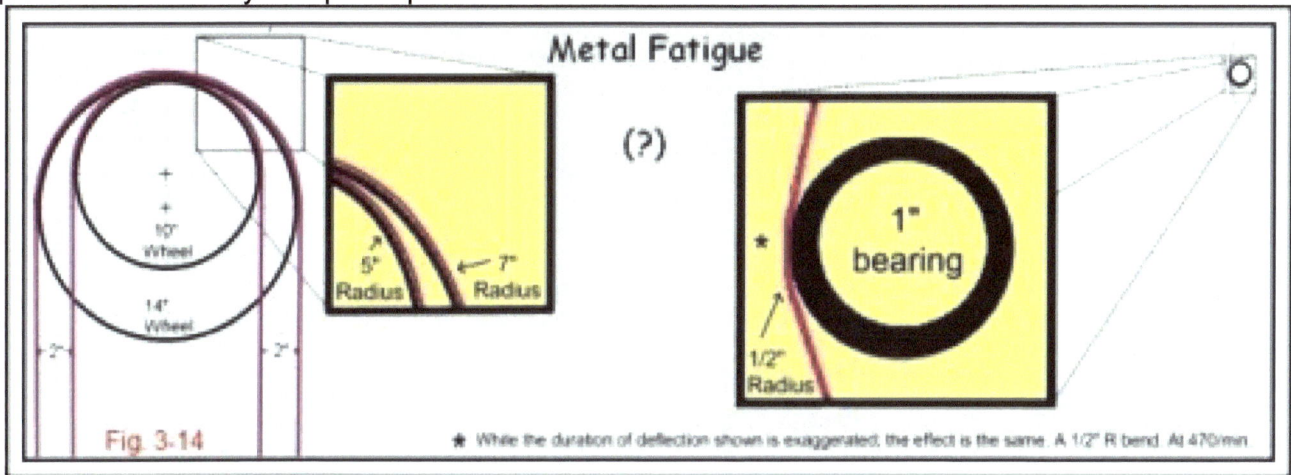

Metal Fatigue

(?)

10"
Wheel

14"
Wheel

8°
Radius

7°
Radius

1"
bearing

1/2"
Radius

Fig. 3-14

★ While the duration of deflection shown is exaggerated; the effect is the same. A 1/2" R bend. At 470/min.

The most frequent cause of blade breakage is deflection from rapidly changing platen angles relative to the kerf when the blade is trying to re-align itself with the lower wheel. During these off-course excursions the blade exits the bottom of the workpiece off-center to the wheels' nominal alignment and forces the blade around a 1/2" radius bend as it re-enters the lower blade guide at 470 blade revolutions per minute. This is how you break welds:

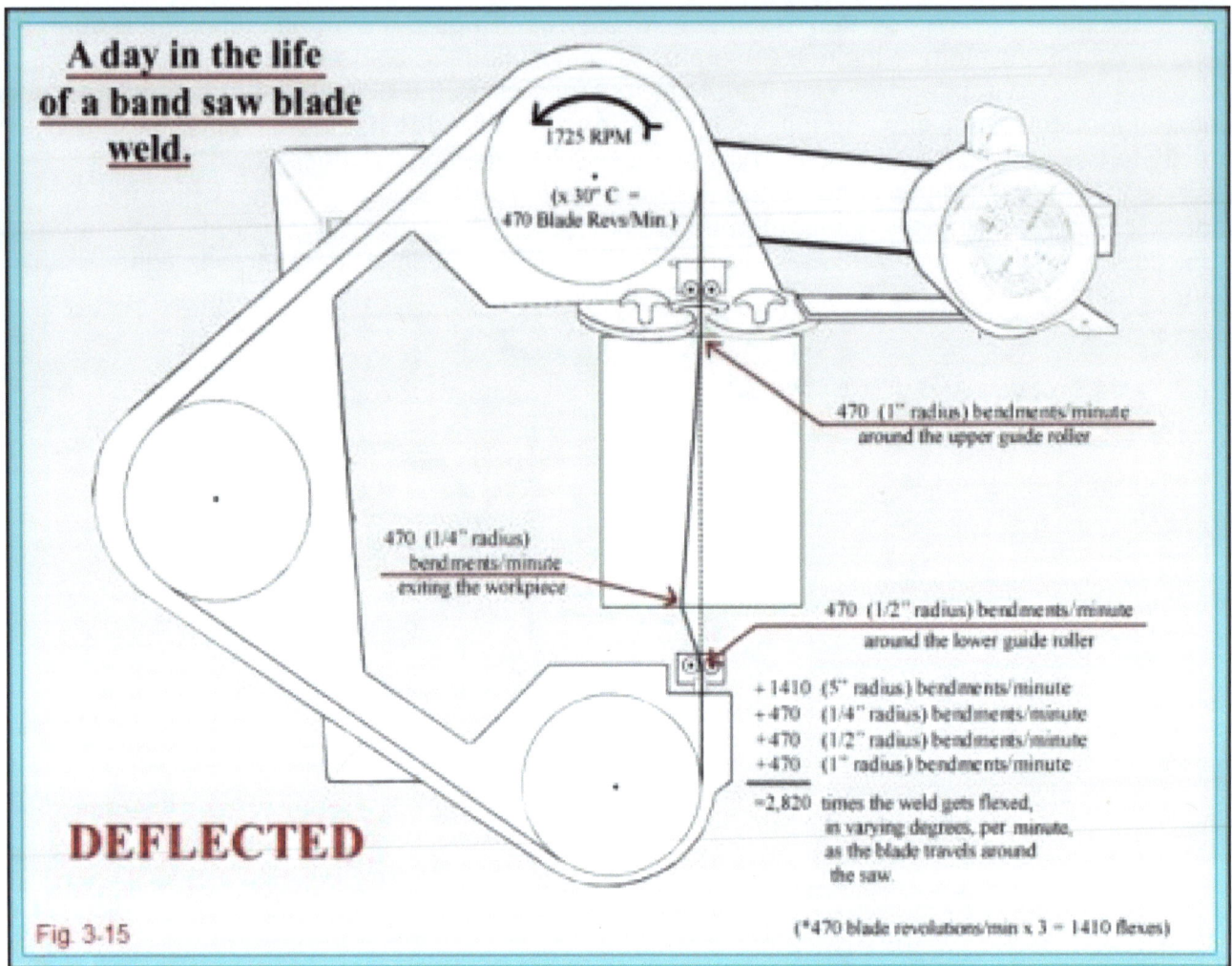

A day in the life of a band saw blade weld.

1725 RPM

(x 30" C = 470 Blade Revs/Min.)

470 (1" radius) bendments/minute around the upper guide roller

470 (1/4" radius) bendments/minute exiting the workpiece

470 (1/2" radius) bendments/minute around the lower guide roller

+ 1410 (5" radius) bendments/minute
+ 470 (1/4" radius) bendments/minute
+ 470 (1/2" radius) bendments/minute
+ 470 (1" radius) bendments/minute

= 2,820 times the weld gets flexed, in varying degrees, per minute, as the blade travels around the saw.

DEFLECTED

Fig. 3-15

(*470 blade revolutions/min x 3 = 1410 flexes)

Three wheeled saws have a bad reputation for breaking blade welds which is grossly undeserved for this reason. It's not the radius of the wheels that causes shorter blade life; it's the jobs we ask three-wheelers to perform. We use three-wheelers to handle bigger timber; that's the main advantage to the third wheel: bigger throats. Timbers are usually pretty rough and when you leave a timber to dry for a couple days you'll find the knots and splinters already raising themselves up off a newly planed surface. You can minimize the damage somewhat by understanding the geometry involved with blade deflections but there's no getting around the fact that working timber is tough business and the blade is the first thing to go. Timber framers charge accordingly and it typically cost $50 per linear foot to rip it and $100 or more for rafter tails or corbel ends. Looking at the turn radii a *blade* undergoes while the saw is cutting a timber, as shown below, you can see that the 10" wheel diameter isn't even close to being the most stressful part of the blade's transit cycle. If you were to bend a blade through a 1/4" radius with a pair of pliers on either side of the weld at the rate of 470 times per minute, which you can't, I guarantee the weld will pop in no time at all. If you wrap the blade around either blade guide roller (3/8"R) by *leaning* against it, or if your blade exits the bottom of your 16" timber out-of-line (1/4"R) you're bending it pretty hard; and 10" wheels are the least of your problems.

When Delta discontinued sales of their 8" wheels my fledgling business was thrown into a total tailspin from which I didn't think I'd recover. I searched the internet high and low for either 8", 9", or 10" bandsaw wheels. You'd think, with all the various bandsaws on the market, there would be somebody making such a basic part but after months of searching I found nothing but cheap, skinny wheels with 1/2" rim widths. There was no way they could be made to track in the manner described above. If anything, I would have liked to find wheels with a 1.5" or even 2" crowned rims. I tried looking for any other kind of wheel that could be modified to act as a band saw wheel but had no luck there, either. It looked like the only way I was going to stay in business at all was to start making my own wheels:

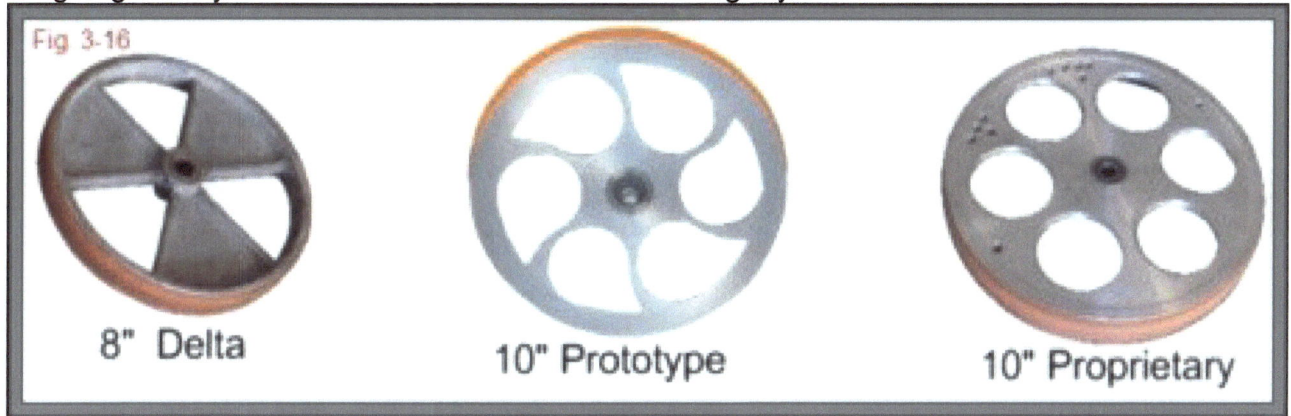

Fig 3-16

8" Delta   10" Prototype   10" Proprietary

Between 2000-2003 I had been making four different-sized saws with 12", 14", 16", and 18" cut depths; all having the same throat width - 18". Using 8" wheels I could shape the frame to fit the shorter cut depths but when I converted to the 10" wheels the frame became *way* disproportional for the smaller saws. The 10" wheels did tuck nicely into the 18" deep throat , however, and fit perfectly into the triangular shape of the new skeletal frame design, which I was also trying to incorporate at that time. The skeletal frame was a concept without a means of producing it. In keeping with the weight restrictions required for the saw to remain portable, it was necessary that all non-essential wheel and blade cover functions be minimized while the most essential blade transport functions were integrated into the frame. Up until then the frame was laid out as a single 1/4" aluminum plate which served as the wheels' backplates and onto which I could lay out the wheels' centers with a T-square. Laying out those centers in abstract space for the skeletal design would require that I construct a very precise jig with the hub co-ordinates pre-defined. The matching one-piece safety cover which I'd also need (to keep weight down) would then have to be cut on a water-jet and welded to fit a corresponding jig. It took considerable time and effort to develop such a precisely mated frame and cover jig so I settled on the 18" X 18" Titan throat; dropping the 12", 14", and 16" models permanently.

No book on bandsaws would be complete without a thorough discussion of the pros and cons of cast iron band saw wheels. Heavy cast iron wheels obviously have a distinct advantage in providing the *flywheel effect* so desirable in a *serious* re-saw band saw. Sometimes it's just not enough to have a 5HP, 3-phase motor and you need a little more oomph to get you through the knots. Presuming you have enough power to start them rolling, that extra inertia really smooths things out when you're force-feeding a 18" x 18" ironwood log through the mill. I find them especially useful on my steam-powered custom band saws. This is a special application band saw that many woodworkers probably aren't aware of but it's really handy during power outages. Our wood-fired steam saws are used throughout the world. They are very popular in the emerging craft-based economies because they burn whatever wood scrap they produce for a clean, green workplace. If anyone says there's no earthly use for a cast iron wheel they haven't seen a steam-saw. You can see, just looking at it, that the toggle flipper would bind in the swing-rod slide cage if the rotational continuator wasn't tied to a heavy-duty flywheel.

Fig 3-17

Otherwise, on the con side of the argument, flywheel-sized blade transport wheels have no use whatever on electric band saws because they require more power to start them rolling than they'll ever use in the normal course of cutting wood: especially when your saw's throat has less than 16" of cut depth. Many woodworkers seem to think a low, earth-shaking rumble is quieter than an aluminum-wheeled, modern band saw and they have a right to their opinion; but if you really want a quiet band saw you'll have to dynamically balance every moving part and that's pretty expensive.

Getting back to the saga of the Falberg wheel design. Who'd have ever thought making a wheel could be so complicated. The only cost-effective way to make such wheels was to cast them in aluminum and machine the bearing pockets and rims. The axles had to be at least 5/8" D in order to seat solidly perpendicular against the relatively soft aluminum backplates without digging in and working loose. The wheel hubs, therefore, had to be big enough to seat a radial ball bearing with a 5/8" ID , the wheel diameter had to be 10" (obviously), and the wheels' width had to hold a 1"W urethane tire with a 1/16" outer ridge to hold the tire in place (because I got very tired of gluing tires on). The only part of the wheel's design in which I had a choice was the spoke pattern, so I jazzed it up with some swirling teardrop-looking cut-outs that made balancing nearly impossible. So, in this case, I screwed up the only design decision that was really arbitrary. All right; I'm sorry!

I called a few places to get quotes on dynamically balancing my wheels and they all wanted more than $100 each; too rich for my blood. The second-best thing was to statically balance them so I had my machinist whip up a slick little spindle thing that hangs a wheel by its center-point and drops the center of gravity below its point of equilibrium.

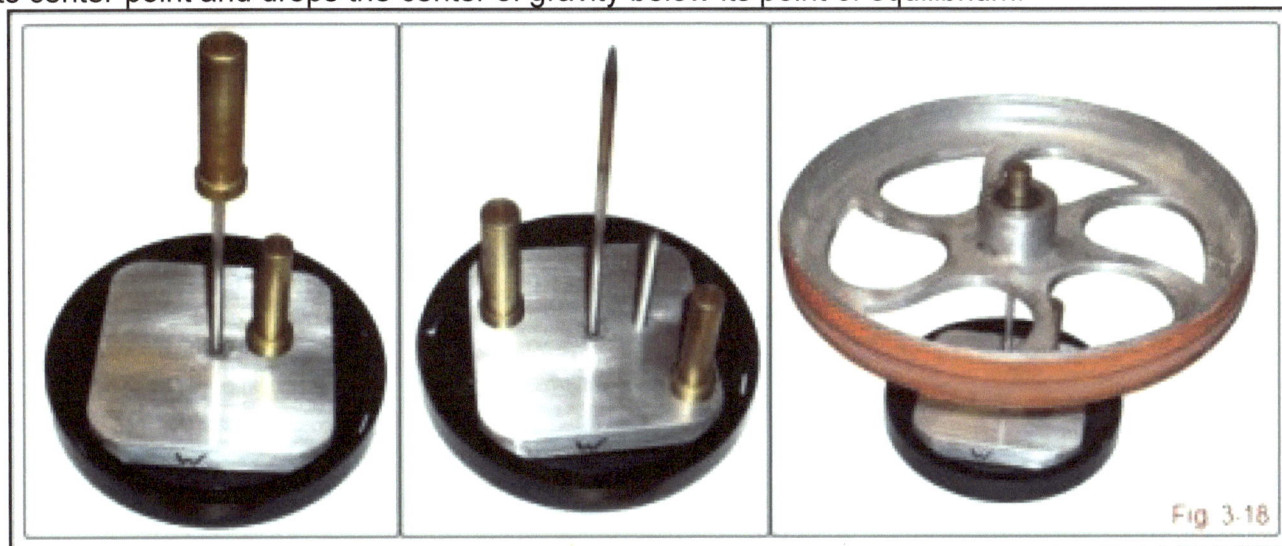

Fig 3-18

You can make any number of variations on this simple jig to reduce vibration on your wheels and pulleys. These brass fittings have a nicely centered cone inside the drilled end and the vertical spike is 3/16" drill shaft with a pointy tip. The wheel will settle into a horizontal plane when it's balanced. I have a 1/2" brass adapter for the idler wheels and a 5/8" adapter for the drive wheels. The short pin is just to hold the spare fitting when not in use. (I wouldn't want to lose it in the shop.) The black plastic base is a CD holder with the post cut off and I store it with the clear plastic cover protecting the balance post, because it's important that you keep it nice and vertical. It's surprisingly accurate and it's quite simple; drill it where it droops.

CHAPTER FOUR
The Magic Formula

The primary function of a Falberg Portable Bandsaw was, and still is, the free-hand cutting of contoured details in large timber. Falberg portable bandsaws fill a unique niche in the woodworking industry and until now there hasn't been much information available about the art and science of cutting through 18" of timber (that doesn't want to be cut) with a hand held band saw. There has never been a blade designed specifically for the kind of work my saws were designed to do. If you shop around for woodcutting bandsaw blades of 1/2" or less, you'll find their specifications nebulous at best. I attribute this lack of specificity to a general unwillingness on the part of manufacturers and consumers to address the complexities of blade dynamics. There are so many minute but significant variables involved that the majority of us would rather discuss blade dynamics in gross generalities at best and brand names at worst. It is my hope in this chapter that by slightly complicating the description of a band saw blade I can greatly simplify the process of blade selection. As much as I'd like to say "Buy Brand X blade, it will solve all your bandsaw problems." I can't. There are too many variables involved. Getting the right **blade** (3) on the right **machine** (2) for a certain **job** (1) takes in three distinct *sets* of variables that have to work together proportionally to formulate what is called a **cutting solution** (4) and it's not as mystifying as some would have it.

(1) **Job**. Obviously the workpiece or job is our first consideration in any cutting solution. The blade/machine combination will depend on the user's expectations, budget, and preferences. Is your minimum requirement to rip 18" wide ironwood veneer or will you occasionally cut 3/4" strips of balsa; there's a huge difference, both in the machine's minimum size and the blade's maximum size; they're inter-related. There are several defining questions:
   a. How fast do we want to cut it?
   b. How smooth does it have to be?
   c. Is the grain straight?
   d. How much sap is still flowing through it?
   e. How hard is the wood?
   f. What species?
   g. How dry?
   h. How deep is the cut?
   i. Are we cutting the same thing all day every day in a production environment, or once-a-month odd jobs?

(2) **Machine**. Any dime-store bench-top can cut 3/4" strips of balsa but band saws get heavier, pricier, and harder to set up as the workpiece and job requirements get bigger, faster, harder, wetter, and more frequent. Each of these variables are inter-dependent and form a mechanical chain of blade control that's only as strong as its weakest link. The weakest link will determine how much blade it can properly support and how much job it can do. Band saw machine variability factors are almost too numerous to list here, but I'll mention a few, for illustration. There's:
   a. Frame size, throat height and width, rigidity, and construction;
   b. Wheel diameter, width, crowning, composition, tires, and adjust-ability;
   c. Motor horsepower, gearing, and wiring;
   d. Blade guides, style, adjustability, and mountings; and then there's:
   e. Table, its trunnions, fence, and whatever in-feed and out-feed accessories are attached.

(3) **Blade** is the most important part of a bandsaw and the least understood. Blades come in more sizes, shapes, and colors than band saws but are described in the fewest and vaguest of terms; usually width and pitch, such as 3/8"-6TPI, 1/2"-4TPI. etc. Blade length is another important spec but it's usually of no importance to anybody other than you and your blade supplier. Other than these broad blade specs the brand name is the only description commonly used to identify a blade's qualities. Several other specifications of equal or greater importance, either by design or ignorance, are seldom mentioned in blade suppliers' literature and the most important spec of all is totally omitted. Blades, too, have many variables incorporated into their construction and these include:

      a. Tooth Pitch – the number of teeth per inch of length;

      b. Thickness – in thousandths of an inch, the thickness of the steel band;

      c. Width – from the tips of the teeth to the back edge;

      d. Composition – the steel alloy of which the blade is forged;

      e. Tooth composition – of which some are made from carbide steel in a separate process and welded on later;

      f. Kerf – the width of a blade's cut, based on the side-to-side spread of its teeth; and

      g. Set Angle – an expression of the difference between a blade's kerf and its thickness.

There are many other blade variables but you get the idea and rather than throw a bunch of brand names and hazy generalizations against the wall at this point I'd like to introduce a logical approach to cutting solutions that we can all agree to, profit from, and use to share our combined wisdom across the lines of brand loyalty, theoretical perceptions, and on-the-job experiences. Until we identify the critical elements of a cutting solution and correctly quantify their effects we'll forever be comparing apples to oranges, to the benefit of nobody.

The bandsaw blade itself is at the heart of almost every contradiction relative to cutting solutions. The blade is the equalizer that enables cheap 14" saws to perform perfect cuts in tough workpieces as well as an expensive 24" saw when the operator understands the dynamics involved. It's primarily what separates the skilled operator from the frustrated beginner.

One-Trick-Pony bandsawyers are but one source of confusion in the matter of blade choices because they stumble, by luck or experimentation, onto a cutting solution that works for their particular applications and preferences without knowing the reasons why. That's great, and we all profit by sharing that sort of information amongst ourselves to advance the craft of woodworking but problems arise when such cutting solutions are offered without reference to the full range of pertinent variables involved. Or, as Paul Harvey would say, *the rest of the story.*

Complicating matters further, the manufacturers and distributors of band saw blades either don't know or won't publish complete specifications about their products. It's often said that the bandsaw is the most difficult power tool in the woodworker's shop to master and it's no wonder we're all confused when all we get is false, misleading, incomplete, or incomprehensible information from the people who purport to be experts. It's understandable that corporate sales executives would disseminate fuzzy claims but supposedly un-biased authors and editors do it too because they just don't know what they're talking about and cherry-pick what few cutting solutions they've succeeded at to present as examples of a broader knowledge that isn't really there.

If you've gotten to the point where you're reading this book to make sense of all the contradictory information you've already accumulated on the subject, you know what I'm saying. Bandsawyering is a very complex subject involving innumerable variables. There is never a simple answer to a simple question. The best I can hope to achieve with this chapter is to acquaint the reader with a full description of the parameters, organize them into a coherent list, assign some priorities, and differentiate between what we know, what we think we know, and what we don't know.

For example: it's often said, generally accepted, and frequently published by people who should know better, that the narrower a blade is, the tighter it will turn a radius. That's the kind of simple statement that makes me scream because it totally ignores the element of blade set. It's only true if you're talking about blades of equal set, and not all blades are set to the same degree. Depending on a blade's pitch and gullet depth, set can vary greatly between blades of the same width and make such statements absurd.

Turning On A Dime

You might recognize the picture on the right from our video. The trick to tight turns is to stay off the thrust support bearing and concentrate on twisting the blade between the side rollers, keeping the blade in a straight line between the top wheel and the scribe line. You have to continually slide the saw sideways, toward the center of the plug, to keep the blade from becoming deflected. This requires looking at three things at one time: the wheel, the guides, and the scribe line; to keep them lined up. It keeps you awake, but it's not that hard. Wear ear-muffs.

I've re-set 1/2" - 2TPI blades to a kerf width of .128" and cut 1/2" radius plugs out of 15" deep Doug Fir despite all published indications that a half-inch blade's minimum turn radius is no less than 1.25".

Narrower is not necessarily Tighter: Wider Is

W (Nom.)"	W (Dec.)"	T (set)"	TPI	K (max.)"	SA °	MTR "
1/8	.125	.014 x3	10	.042 N/A	~~6.39~~	~~.147~~
1/8	.125	.014 x2	14	.028 ?	3.20	.283
1/8	.125	.014 +4	18	.018	0.92	.978
1/4	.25	.018 x3	6	.054 ?	4.12	.443
1/4	.25	.018 x2	10	.036	2.06	.873
1/4	.25	.018 +6	14	.024	0.69	2.606
3/8	.375	.022 x3	3	.066 ?	3.36	.810
3/8	.375	.022 x2	6	.044	1.68	1.604
3/8	.375	.022 +8	10	.030	0.61	4.397
1/2	.5	.030 x3	2	.090	3.43	1.057
1/2	.5	.030 x2	3	.060	1.72	2.091
1/2	.5	.030 +12	6	.042	0.69	5.211
5/8	.625	.036 x3	2	.108	3.30	1.374
5/8	.625	.036 x2	4	.072	1.65	2.722
5/8	.625	.036 +12	6	.048	0.55	8.141
3/4	.75	.040 x3	1.5	.120	3.05	1.778
3/4	.75	.040 x2	2	.080	1.53	3.526
3/4	.75	.040 +12	4	.052	0.46	11.721
7/8	.875	.042 x3	1.5	.126	2.75	2.299
7/8	.875	.042 x2	2	.084	1.38	4.568
7/8	.875	.042 +12	3	.054	0.39	15.954
1	1	.042 x3	1.5	.126	2.41	2.997
1	1	.042 x2	2	.084	1.20	5.963
1	1	.042 +12	3	.054	0.34	20.836

Fig. 4-02

N/A because while it's mathematically possible it's not technically feasible
? : because I doubt one could actually get that much set at that scale

The chart above is a mathematical model based on hypothetical variations of blade set applied to the range of blades common to the vertical woodworking bandsaw. The Minimum Turn Radii shown in the red squares illustrate the tremendous difference in maneuverability between blades designed for re-sawing and blades designed for contour work. Nominal blade Width, shown in groups of three, are calculated from three possible set variations. The first, noted in red as (x3), represents a typical blade of coarse raker pitch re-set to its maximum Kerf, given a proportionate Thickness for its Width. The second entry in the set of three represents a finer alternate-set pitch re-set to twice its Thickness. The third entry is typical of the set you'd find on an off-the-shelf blade and sold as a resaw blade.

Relative to the "Turning On A Dime" illustration above: I cheated and used a re-set blade specifically set for that cut and most woodworkers won't go through that much effort to re-set a blade by hand. I have a machine of my own design to do tricks like that but there is also a substantial variation in the blade sets (K) of commercially available blades and they often cross the lines of manufacturers' stated radius values to out-turn blades of greater width (W). One *could* say that blades of greater set angle (SA) turn tighter radii, but that would be false, too; blade width (W) does play some part, it's just not the primary factor.

Complicating the matter further; the narrower (W) the blade is, the less one CAN set the teeth. A tooth that only rises 1/64" above its gullet can only be set to 1/32" (K), and that would require setting the tooth sideways at a 90 degree angle; which can't be done. See how misleading it is to say a narrower blade turns tighter radii? There are other variables that affect minimum turn radius (MTR) but first among them is the relationship between blade width (W) and blade set (K), expressed as an angle; the set angle (SA).

I measured kerf width (K) by sandwiching the off-set teeth between two short sections of blade stock (with the teeth ground off). Using a micrometer I measured the entire sandwich, then subtracted the thickness of both blade sections to arrive at the total set (K). Blade thickness (T), if it's not already listed in the blade specs, is easy enough to do yourself with the same micrometer. Then it's a relatively simple matter to plug those numbers into the formula below to *approximate* set angle (SA). I say approximate because real scientists would call it sloppy science, but it's close enough for wood work. I allowed for 10% error.

Where: $K=$_____ $T=$_____ $W=$_____ $\&\ X=$

$SA = ((\underline{\ K\ } - \underline{\ T\ }) \div 2) \div ((3.1416 \times 2 \times \underline{\ W\ }) \div 360)$

$SA = ((\underline{\quad} - \underline{\quad}) \div 2) \div ((3.1416 \times 2 \times \underline{\quad}) \div 360)$

$SA = \underline{\qquad}$

$TR = (\underline{\ W\ }^2 \div (8(\underline{\ K\ } - \underline{\ T}) \div 2)) + (((\underline{\ K\ } - \underline{\ T\ }) \div 2) \div 2)$

$TR = (\underline{\quad}^2 \div (8(\underline{\quad} - \underline{\quad}) \div 2)) + (((\underline{\quad} - \underline{\quad}) \div 2) \div 2)$

$TR = \underline{\qquad}$

Fig 4.03

Fig. 4-04

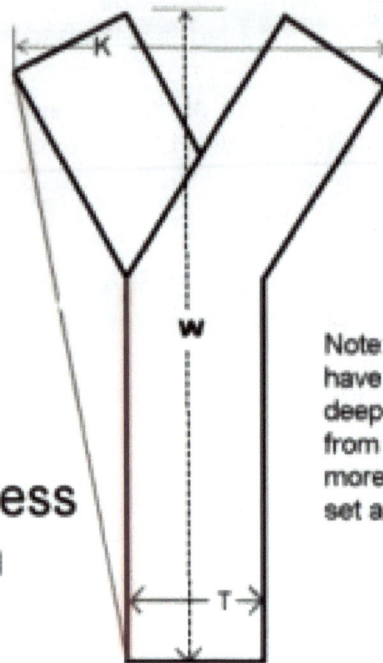

Where:
$\Pi = 3.14$
K=kerf width
T=blade thickness
W=blade width
SA=set angle

Note: Carbide-tipped blades have a rectangular tip 1/16" deep so subtract that 1/16" from blade width to get a more accurate measure of set angle to back of blade.

$$SA = \frac{K - T}{2} \div \frac{\Pi \cdot 2 \cdot W}{360}$$

A 1/2" -2TPI blade re-set to .128 K would work out thusly:

$$\frac{.128 - .032}{2} \div \frac{3.1416 \cdot 2 \cdot .500}{360} = .048 \div .0087266 = \underline{5.5}\,°$$

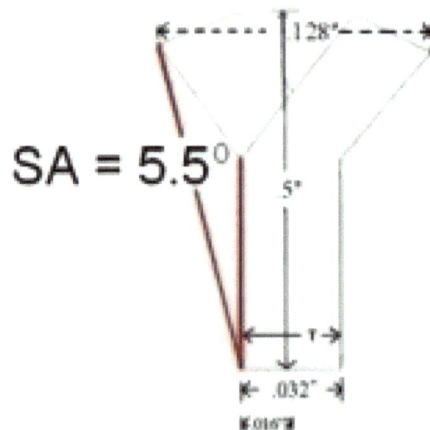

$SA = 5.5°$

128"

5°

.032"

.016"

Among other things, set angle (SA) tells us what a blade's turn radius (MTR) will be, relative to its width (W). Given two blades with equal set angles the narrower blade **will** turn a tighter radius. Set angle determines how well a blade will resist workpiece deflection and how well it will follow a mis-aligned fence. Likewise, set angle determines how well a blade will resist bowing inside a kerf. Set angle will determine how rough the finish is likely to be.

Indirectly, set angle will determine how much tension your saw will have to support; blades with higher set angles will cut straight through much more wood with much less tension; if you don't mind a rough finish. Of course finish is a big concern but it's not your primary concern; straightness and flatness of cut are your first priority; finish is secondary. While straightness and flatness would lead you to choose a higher set angle; smoothness is best achieved with lower set angles requiring higher tension. Finding a happy balance between these conflicting goals is, therefore, the key to a successful cutting solution keyed to the amount of tension your saw can support. Simple, huh?

The most important relationship in the entire complex of band saw relationships is the least understood because: either bandsawyers, distributors, manufacturers, and writers are either totally intimidated by the formula that defines it; or the remote possibility that nobody's thought of it before. In my desperate search for some rationale behind the behavior of band saw blades I spent a week reviewing my high school algebra to come up with this solution to the problem of blade drift. The math is not scientific but I'm assured by a genuine rocket scientist that it's "close enough for your purposes". It may seem like more work than it's worth to calculate these things but when you see how much it saves you in wasted wood and time by being able to predict the behavior of a bandsaw blade under varying circumstances you'll agree. You won't have to do it very often to get the idea why set angle plays such a vital role in successful operation of a bandsaw. You don't necessarily have to get all scientific and measure each blade's set angle with micrometers. While it's hard to see with your eyes, you can usually *feel* set with your fingers by running them down the blade (backwards from the direction they cut, lest you cut yourself).

As complicated as all the conflicting information and erroneous opinions would lead one to believe, there are "if - then" logical reasons for every aspect of a blade's geometry that, once understood, make blade selection a fairly simple and straight-forward process. While I appreciate simplicity as much as the next guy, I find it extremely difficult to make informed choices when the information I'm given is incomplete or incorrect. The following chart was gathered from what little information the selected manufacturers and their distributors choose to publish on-line. You might notice that there are several blade distributors not mentioned and it's not because I forgot them; it's because they offer no specifications at all, other than the length and, of course, the price.

There are several interesting discrepancies between the authors of these numbers but, over-all, they tend to agree in the broadest terms and based on my own experience I'd have to add that they seem to apply mostly to cut depths of no more than 2" and are therefore useless to somebody working with big chunks of hard wood. The nature of the workpiece isn't even mentioned in their reporting so we're left to assume that it doesn't matter; **but it does**. The numbers I got didn't agree with the numbers they got mostly because I was using re-set blades on Doug Fir 16" thick. It's not important except to show the contradictory nature of what little information is available, and how poorly we define the basic question: "What blade should I buy?" There is a correct and appropriate answer; but not a simple answer. I'm going to assume the reader, that being you, is capable of sorting out the details for yourselves.

The blades listed above are just a small sampling of the blades in use today and someday I'd like to compile a complete list of every blade out there with their specifications listed in the same ascending order of set angle. Blade distributors won't like it, as I'm sure they'd rather supply all of your bandsawing needs by promoting brand loyalty without getting too specific about the facts. The truth is, you're going to have to shop around and keep a small sampling of several brands' offerings to fill all the niches in your quiver of cutting solution *arrows*. When you're shopping around you should demand to know the set angle of a blade before you commit to purchasing it. Their labeling of blades with brand and model names like "The new ACME **"Ajax Lightning""** have absolutely no meaning when you're buying a commodity like saw blades. You need complete specifications to make an intelligent choice and they are failing their customers miserably by not providing it. Stand up! Storm the Bastille!Demand answers!

Available Blade Specs

		Width	Tooth Per Inch	Thickness	Kerf (width)	Set Angle	Minimum Turn Radius	NOTES
Suffolk Mach. Corp. - TW		1/8	14	025	?	?	21875	
		3/16	4	025	?	?	375	
		1/4	4	025			625	
		3/8	2	032	128	7.33	5625	
		3/8	3	032	064	2.44	8125	
		1/2	2	032	064	1.83	2.5	
		1/2	2	032	090	3.32	1.09	ReSet
		1/2	2	032	128	5.50	750	ReSet
		1/2	3	032	064	1.83	2.5	
		3/4	2	032	064	1.22	5.4375	
		1	2	035	?	?	7.125	
		1	2/3	035	060	0.72	7.125	Carbide-Tipped
Lenox	Diemaster	1/4	6	035	?	?	?	
		3/8	4	035	?	?	?	
		1/2	3	035				
	Tri-Master	3/8	3	032	064	2.44	625	Carbide-Tipped
		1/2	3	025	062	2.11	1.38	Carbide-Tipped
		3/4	3	035	?	?	?	Carbide-Tipped
		1	3	035	?	?	?	Carbide-Tipped
	WoodMaster	1	2	035	078	1.23	5.82	
	"	1/2	2	032	125	5.33	8125	ReSet
Highland	Wood Slicer	1/2	3	022	031	0.52	6.95	
	"	3/4	3	022	031	0.34	15.63	
Laguna	Resaw King	3/4	3	032	060	1.07	5.03	
	"	1	3	032	060	0.80	8.94	
	Shear Force	5/8	?	?				
Olson	Flex-Back	1/8	14	025	?	?	?	
		3/16	10	025				
		1/4	6	020				
		3/8	4	020				
		1/2	3	020				
Carter	Accuright	1/8	14		?			
		3/16	10					
		1/4	4					
		3/8	4					
		1/2	3	026	062	2.06	1.75	

*Data supplied by Falberg Saw Co. is shown in purple

Fig 4-05

I read somewhere that the best way to calculate the minimum turn radius of a fire truck was to dis-assemble the chassis, carefully measure all the parts, re-assemble it, soak the wheels in wet white paint, drive around in a circle, and measure the diameter of the white line. Ultimately it's going to be the easiest way to determine a blades' minimum turn radius also. The technique I use to test for minimum turn radius is to twist the body of the blade between the side rollers and try to spin the saw without touching the thrust support bearing. While doing this I take care to keep the blade in line with the wheels and the workpiece in line with the blade. While it may sound counter-intuitive, the blade **will** advance through the cut. When done with a stationary saw it's important to let loose of the workpiece frequently to re-center the blade in the workpiece and let the blade re-center itself between the wheels; or you're likely to get a bowed cut. When done with a portable bandsaw it's a little trickier because you have to physically move the saw from side-to-side to maintain that three-part alignment.

There are so many other generalities being thrown around that are generally inaccurate that one generally doesn't know where to begin. The notion that wider blades are better for re-sawing while narrower blades are better for contour cuts is so over-simplified that it must be considered totally invalid, even as a generality. The exceptions to the rule are so numerous that merely suggesting it has any relevance does more damage than good for the beginning bandsawyer who is just looking for a starting point in his quest for the mastery of this art.

The first thing, to my way of thinking, that a novice bandsawyer should learn about the over-all set-up and operation of a band saw is the nature of its most important component, the blade. The blade determines how much wood you can cut and how much band saw you'll need to cut it. The blade, with all its variations and complexities, must be chosen with specific objectives in mind; and to do that one must be aware of the specific effects of each feature. Every aspect of a bandsaw blade has a reason for being. The designers of bandsaw blades had something in mind when they decided to use .025" steel for its body. They had something in mind when they divided each inch into 14 teeth. They had specific workpieces in mind when they decided to use a hook-shaped gullet or raker pitch. There's a specific blade for almost every job requirement but unless you understand the reasoning behind each feature and its relationship to the others you'll be guessing, and probably wrongly, every time you choose a blade.

Looking at the beginner's goal, to cut something with a bandsaw, I think it would be more instructive to approach the cutting solution in its logical sequence and address the multitude of variables in the order of their appearance. So the first question is: "What do you want to cut?" If the answer is; "Tiny reindeer out of 2" x 4" lumber", the solution would be to use just about any bandsaw that can track a blade and choose a 1/8" or 3/16" fine-tooth blade with as much set as you can find. As your expectations rise and you decide to strive for a smoother finish you might try making the same curves with a blade of narrower set. Getting even more ambitious, you decide to make big reindeer out of 6" x 6" oak beams. With that depth-of-cut and hardness of material you're going to need bigger gullets to remove the chips, fewer teeth per inch to leave a space between them, a wider blade to deepen those gullets, and more tooth-set/wider kerf to turn the wider blade through tight radii. If your original dime-store saw can't tension a wider blade you'll have to get a bigger saw with a bigger spring and a stronger frame. And on and on and on until you're making payments on a 42" Tannewitz.

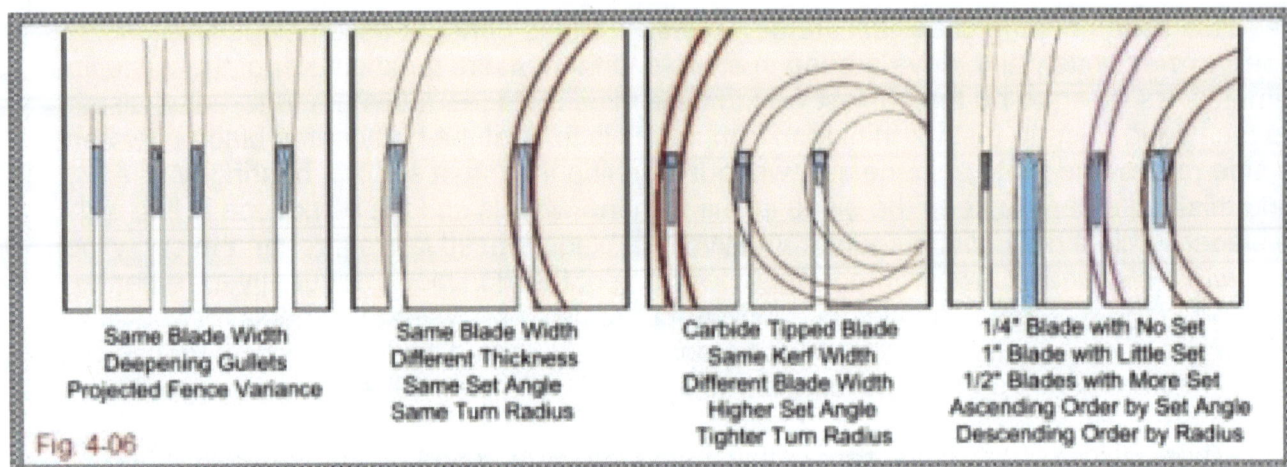

Same Blade Width	Same Blade Width	Carbide Tipped Blade	1/4" Blade with No Set
Deepening Gullets	Different Thickness	Same Kerf Width	1" Blade with Little Set
Projected Fence Variance	Same Set Angle	Different Blade Width	1/2" Blades with More Set
	Same Turn Radius	Higher Set Angle	Ascending Order by Set Angle
		Tighter Turn Radius	Descending Order by Radius

Fig. 4-06

It's easy to see how a blade's set angle relates more to turn radius than a blade's width does but not so easy to see is the relationship between set angle and "blade drift". I parenthesized "blade drift" because it doesn't exist in the real world. The term is commonly used to describe the discrepancy between blade alignment and fence alignment when the two are out-of-sync with each other. If you look at the first drawing of the illustration above, where the "cone of uncertainty" (in hurricane prediction parlance) is projected forward for a blade with no set, two blades of minimal set, and a blade with sufficient set to be more forgiving of blade/fence discrepancies. While the first satisfies our desire for narrow-kerf / smooth-finish veneer cuts, it is extremely difficult to achieve and requires a high level of operator skill, meticulous fence alignment, and a saw capable of very high tensioning. In some cases, even the foregoing prerequisites aren't enough to overcome workpiece deflection from knots and cross-grain variations in figured hardwoods; in these cases your **only** recourse is to increase the amount of set.

Much good wood is lost to the planer or sander in striving for that perfect thin-kerf/ smooth -finish cut and the ambitious bandsawyer flirts with disaster in trying. Let me be first to say "and there's nothing wrong with that", because that's what makes us better craftsmen. We learn our saws' limits that way. I'm just saying that there's an economy of effort to consider as well as whatever concerns the sawyer might have about risking a valuable piece of wood to the ravages of bowing. One can compensate for low set angles by adding more tension; but if your saw has no more tension to offer, wider set is the next option. Conversely; one can compensate for a weaker, less expensive band saw by choosing blades with more set angle. There's an important economy to consider in this regard also, in that narrower blades are less expensive and have the same number of teeth to wear out in the same amount of cutting time. Add it up: cheaper machine, less footprint, easier transportation/ shipping, and cheaper blades, hmmm. How many *thousandths of an inch* of precious hardwood is that worth?

I'm not saying that big, powerful saws aren't worth the money either. There's another factor that enters into the cutting solution and that's feed rate. High-tension machines that are capable of running one inch, narrow-kerf blades enjoy the overwhelming advantage of being able to cut big chunks of lumber very fast, very smoothly, and with nice thin kerfs. In a production wood shop that has a distinct value. But you have to remember; it comes at a cost. Is it economical for you? If you're a hobbyist or mixed-use bandsawyer I'd say no, you're better off with a lighter, cheaper saw and narrower blades. If you use it regularly to cut veneer or re-size hardwoods I'd say get the industrial machine and fine-tune it for resawing with thin-kerf blades. Also: the narrower the kerf, the faster a blade will cut; because it's cutting less wood and creating less sawdust.

It's not magic and there's nothing about band saw blades that can't be explained, measured, and analyzed to predict how they'll perform when pushed through a piece of wood. For instance: if you have a blade with no set it will cut perfectly straight. Whether the fence is in alignment with the blade is another question, but the blade will be controlling the path of the cut. If the same blade encounters a hard knot inside the wood it might try to deflect side-to side to avoid it, creating latitudinal cupping, because sometimes you just can't apply enough tension to overcome the blade's intrinsic elasticity and it starts to twist internally. Therefore: resawing thick material with blades lacking some minimal set makes it nearly impossible to get a flat , straight kerf. You'd have to align the plane of your fence exactly parallel to the plane of your blade, which can be quite difficult.

On the other hand; a 1/2" blade, .032" thick with a set of .016" on either side, produces a kerf of .064" and allows .032" of "wiggle room" for the blade to adjust its own path as deflection occurs or to compensate for a fence that is not in perfect alignment with the flat of the blade. You can cut perfectly straight, flat veneers 18" wide with a 1/2HP motor and never adjust for drift, tracking, or tension. And you could cut a 3" circle that same depth. Deep rough cuts are quick and dirty and get you to the planer in style. Planers are great but if you have to plane away 1/8" of good wood to get flatness you're probably using the wrong blade. You can take a half hour to set up a narrow-kerf cutting solution or you can ram everything through the same wide-kerf blade; that's the trade-off and what you need to consider when buying your bandsaw and the blades you use on it.

Visualize a .032" x 1/2" thin strip of steel flying through 16" of wood at 76'/second. It's going along OK until it runs into a knot and one side of the blade is into harder wood than the other. This creates a situation analogous to an airplane in a cross wind. In order to maintain directional stability it has to crab; crabbing is when you're pointed one way while going another. Airplanes crab into crosswind landings every day without mishap. Now imagine crabbing into a narrow runway with high fences on either side, and your fuselage is too long to squeeze through. The tail hits the fence, the plane straightens out, the cross wind blows it through the fence, and you have a wreck. Likewise; bandsaw blades will crab automatically because the leading edge is stretched between said wheels and confined by their tracking characteristics to a pre-ordained path around the blade transport wheels. Like the plane, if the tail end of the blade bumps into the sides of the kerf it's going to prevent the blade from crabbing, and "spin off into the grass" so to speak. It is, admittedly, a flimsy analogy so if you know a better one, use it. The important thing to understand is that the kerf needs to be wide enough for the flat of the blade to freely follow in the "shadow" of the teeth's kerf without interference. The body of the blade must be free to twist and turn in response to deflectionary forces such as grain patterns, knots, nails, and inconsistencies in the blade's own geometry, such as a distorted, dull, or damaged teeth.

My first twenty saws, even at full spring compression, had barely enough rigidity to support any blade wider than the 3/8" Timberwolfs. I didn't know it then but the reason those blades worked, where no others would, was that they had a set angle of eight degrees. Timberwolf 3/8" - 2TPI blades were short on beam strength because the depth of their gullets didn't allow much blade body to butt weld, so they were somewhat fragile and tended to break when pushed too hard. The now-discontinued 3/8" - 2TPI Timberwolf blades were used for the Bayer posts, but if I had to do it now I would have to use a 1/2" - 2TPI Timberwolf blade.

I had been begging Suffolk Machinery for a 1/2" 2TPI blade since I started this business. They had a 1/2" 3TPI blade which worked nicely enough, but due to its finer pitch, it was slower feeding than I'd have preferred and didn't turn very tight radii. The 1/2" 2TPI I finally did get were, as expected, more robust but left much to be desired in the cutting of tight radii so I had to re-set them by hand to do detailed cuts, which got *old* really fast. This kind of necessity, being the mother of invention, spurred the development of my handy-dandy Set-O-Matic Blade Setter.

A Machine for the Setting of Band Saw Blades

Fig. 4-07

This was my first prototype blade re-setter and it's still in use for 1/2" -2TPI blades. I have it adjusted for .064". The second prototype, below, is convertible from 1/2" (W) to 3/8" (W) and feeds a little faster but still only sets the 2TPI pitch blades. I couldn't get a clear picture of number three ;<) but it sets both 2TPI and 3TPI blades with more precision and less effort. It's of interest here primarily because the manufacturers of these blades said they couldn't be re-set without breaking the hardened teeth. I haven't broken a tooth yet but I've chipped a few by over-tightening the hold-downs. You just have to be careful how you set them up.

Perhaps the most interesting thing about my blade setter project is that it was done entirely with hand tools; no welding, machining, casting, or extrusions required. Using the same design process that produced the portable band saws, I broke it down to one function at a time and started adding pieces to it "brick by brick", so to speak. Whatever parts appear to be machined are really only hand-filed to look that way. The finish was done with 4" polishing disks and some hand polishing with non-woven abrasives, all of which are relatively inexpensive and available at any internet parts supply house like Grainger or MSC. While you're at it, take a look at all the gears, shafts, pulleys, and bearings you can get for cheap to build your own simple machines.

Did I mention that I have no experience or education in industrial engineering whatsoever? I have more fun now building machines than I ever had designing carpentry or woodworking projects. If you've ever enjoyed Tinker Toys, Legos, or Erector Sets you'll love playing with power tools to build jigs, fixtures, and small home-made machines. Designing machines like this requires concentration and if you need something to take your mind off work; this is it! I'll get deeper into basic metal-working techniques later but for now let it suffice to say that such projects aren't nearly as difficult as they might seem.

Fig 4-08

Below is how they did it at the big saw mills and there are even slickers versions available today for lumber mills but they only do 1" or wider blades so they don't apply to our petty little concerns. Since it's not deemed economically productive to re-set blades for vertical woodworking bandsaws there isn't a narrow-band re-setter commercially available so the average Joe is left with whatever set the blade suppliers choose to give us, which isn't much. I thought seriously about patenting and marketing the setters developed here but couldn't find anybody to join in the effort and, lacking the time and resources, dropped it. If anybody is interested I'd be happy to assist. There's no big secret to re-setting blades and my timber-framing customers have done it for years with hammers and punches or with pliers-looking blade-setters like this:

Fig. 4-09

Ye Olde Blade Setter

Having my own blade setter to experiment with gave me a new perspective on blade geometry and I'm not exactly sure what it is, but it's way different from the mainstream of bandsaw blade advice being offered to the general public. My biggest complaint is that blade suppliers simply can't or won't provide the most relevant specifications about the blades they're pushing and the buyer is given no option but to "try this" and "try that". Forcing the consumer to buy a lot of different blades he can't use might help to boost sales for the distributor, it's a major dis-service to the consumer who should be mad as hell and not taking it anymore. It's insulting to be treated like a child that "wouldn't understand" the details and told to just "take it from me, I know what's best for you". They don't.

I suspect the distributors of band saw blades could save themselves thousands of hours of customer service phone time by simply stating their blades' set angle (SA). You frequently see their blades' thickness (T) listed but seldom their kerf-width (K). I may be easier to sell blades that way but it doesn't serve the consumer well and this is something we'll probably have to do for ourselves. If you'd like to enter your favorite blade into this database I'd be happy to include it in future editions. All you need is a set of mechanics' feeler gauges, a micrometer, and a few short pieces of blade stock (any width or thickness).

Consumers, that being you, can salvage a lot of un-used blades by re-setting them, by hand, using a hammer and a punch. Simply clamp the blade into your vise with the gullets flush to the jaws and give the teeth a little whack as close to the base of the tooth as possible. It goes pretty fast once you get the rhythm of it and you'll be surprised how accurate this method really is. And you'll be amazed how different the saw behaves with a wider set. For one thing you'll see how much easier it is to follow a scribe line. You'll also notice the blade cuts considerably slower and creates twice the sawdust as it did before.

There is no such thing as a good *all-purpose* blade. There is no such thing as a *good resaw blade*. There is no such thing as a *good contour blade*, either. *Good blades* **have** to be defined in terms of what you intend to cut and what kind of machine you intend to cut it with. Conversely; there's no such thing as a *bad blade*; for what might be bad qualities for you are exactly the kind of qualities another bandsawyer is looking for. The words *good* and *bad* simply don't apply to band saw blades (with the possible exceptions of price and weld integrity values). Taken in the context of machine and workpiece variations, a blade can only be said to be *appropriate* or *in-appropriate*. One can make valid comparisons and judge one blade superior to another only if the cutting solution, as a whole, is considered. Even then, there are minor qualities that one bandsawyer will find more important than another and such value judgments would still remain undecided.

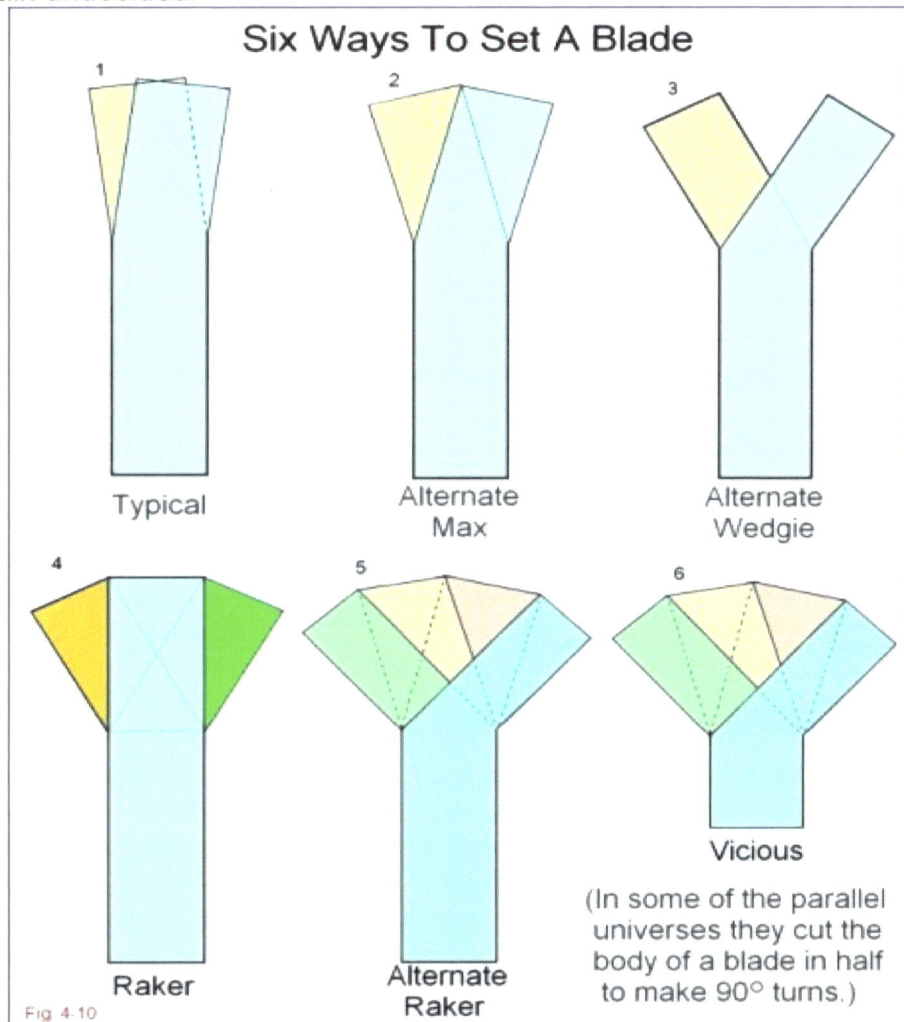

Six Ways To Set A Blade

1 — Typical
2 — Alternate Max
3 — Alternate Wedgie
4 — Raker
5 — Alternate Raker
6 — Vicious

(In some of the parallel universes they cut the body of a blade in half to make 90° turns.)

Fig 4-10

There are several ways of setting a blade to perform specific cuts but they all depend on having enough gullet depth to execute a bend at the base of the tooth. In my zeal to create the widest set ever inflicted on a "stock" blade I deformed a 1/2"-2TPI Timberwolf blade into what I now call the Alternate Wedgie design which works fine if you're not in a hurry and intend only to cut minimum-radius-turns all day, but trying to rip-cut with an Alternate Wedgie (3) set blade is an exercise in futility because the V-shaped notch between the set teeth serves as a very effective brake to any forward motion. It also leaves a very strange kerf with a pointy little ridge down the center when you stop the saw and back up a bit. The extremely wide sets of Alternate Raker (5) & (6) require not just one, but two, subsequent raker teeth to eliminate the gap in coverage, each of which must also be alternately set, to a lesser degree. The Alternate Raker set is twice as time-consuming as the Raker set because it has to be run through the setter twice but probably wouldn't make so much difference if done by hand with a hammer and punch. The Alternate Max (2) is really typical of many standard blades advertised as wide-kerf for general purposes. My favorite is the simple Raker (4) which I tweak to the proportion shown for most corbel cutting applications where deep cuts of radii less than 2" would be encountered. Blades of this kerf width will turn nicely and back out of radius cuts easily enough to maneuver back into a second approach for right angles. Whenever these sets are considered you have to also consider that doubling a blade's set cuts your feed rate in half and making twice as much sawdust.

Fig. 4-11								
Ratio Projection								
Teeth / W	Thickness/ Width	W (Nom.)	W (Dec.)	T	TPI	K (3.5 x T)	SA	MTR
1	.064/1	1/16	.0625	.004	16	.014	5.5	.100
↑	↑	3/32	.09375	.006	10.6	.021	↑	.150
		1/8	.125	.008	8	.028		.200
		3/16	.1875	.012	5.3	.042		.300
		1/4	.25	.016	4	.056		.400
		5/16	.3125	.020	3.2	.070		.500
		3/8	.375	.024	2.6	.084		.600
		1/2	.5	.032	2	.112		.800
		5/8	.625	.040	1.6	.140		1.002
		3/4	.75	.048	1.3	.168		1.202
↓	↓	7/8	.875	.056	1.1	.196	↓	1.402
1	.064/1	1	1	.064	1	.224	5.5	1.603

Just as a casual observation, have you noticed that the most popular contour blades have a blade width (W) to pitch (TPI) ratio of 1:1 while re-saw blades tend to have a width to pitch ratio of 1:2 ?   Between the blade widths of 1/8" to 1", the thickness (T) of blades currently in favor ranges from .014" to .042". Using that as a scale, the median thickness of a 1/2" blade would be .028" which is close to reality, but projecting that same ratio down to 1/16" (W) would skew that blade's thickness down to an impossibly thin foil. To illustrate this I projected a *golden ratio* of 1"/.064" for W/T in both directions throughout the index of nominal blade widths to show the direct relationship between Set Angle and Minimum Turn Radius. I applied the 1:1 ratio of Teeth Per Width to project an indexed TPI value to each width increment from 1/8" (W) to 1" (W) for additional scale. Considering that an aluminum beer can is between .010" and .019" thick, the projected thickness of blades under 3/16" on my imaginary chart would not be physically viable; nor would the projected thickness of blades of more than 3/4" width. To compensate for the machine dynamics associated with running impossible blade thicknesses one would have to introduce an "X" factor which would totally ruin my hypothesis. Curses! I don't know what machining or metallurgical challenges are involved in the production of a 1/16" 16TPI .004"T blade would be but , man, talk about a
"light sword". There's the issue of beam strength and deep cuts could be problematic but on the other hand, those would be some mighty fine detail saws and in terms of feed pressure PSI & feed rate, it would be an interesting experiment.

It would seem like everything having to do with bandsaws involves *buying a pig in a poke*. Deceptive advertising aimed at securing a greater market share by claiming their products are the answer to everyone's needs does a dis-service to all. If they think the consumer isn't interested in the details or wouldn't understand the data, they need a new focus group. Shrouding the specifications of their wares in a fog of mystery wastes our time as well as theirs and nobody profits. How hard is it, really, to refer to blades by their full name? Labeling a blade "3/8-3TPI" tells us nothing about its performance but if you refer to it as a "3/8-3TPI, .028T, .064K" blade, we'd know everything we need and thus be able to make informed decisions on whether the blade is suitable for the job at hand. "Wood Mutt" , "Beam Ripper" , "Hardwood Slicer" , or any other such silly brand names mean nothing and only add to the confusion when what we're looking for are solutions. It would help even more if blade distributors would distribute their blades' Set Angle and Minimum Turn Radius at a given cut-depth. I doubt any of them would lose any sales and they'd all profit from the standpoint of less need for customer service after-the-sale. Brand allegiances may shift as a result, but in the end they'd all do more business with happier woodworkers and get far fewer returned blades.

There are some generalities that **can** *truly* be said of band saw blades:
1. The higher the set angle, the tighter the turn radius.
2. The higher the set angle, the less it can drift.
3. The wider a blade is, the broader its cross-section, and the more it resists thrust.
4. The wider a blade is, the more tension it will require to suppress flutter.
5. The thicker a blade is, the more tension it will require to suppress flutter.
6. The thicker and/or wider the blade, the more resistance it has to feed pressure.
7. The deeper the gullet, the wider the teeth can be set.
8. Carbide-tipped blades, by virtue of their squared teeth, leave a smoother finish.

There are some generalities than **can** *truly* be said of band saws:
1. The stronger the frame, the more spring tension it can support without flexing.
2. The more spring tension, the wider the blades it can support.
3. The wider the blade, the higher the load on the motor.
4. The heavier the wheels, the greater their inertia.
5. The longer the span between wheels, the more tension is required.
6. The heavier the saw, the harder it is to move it around.

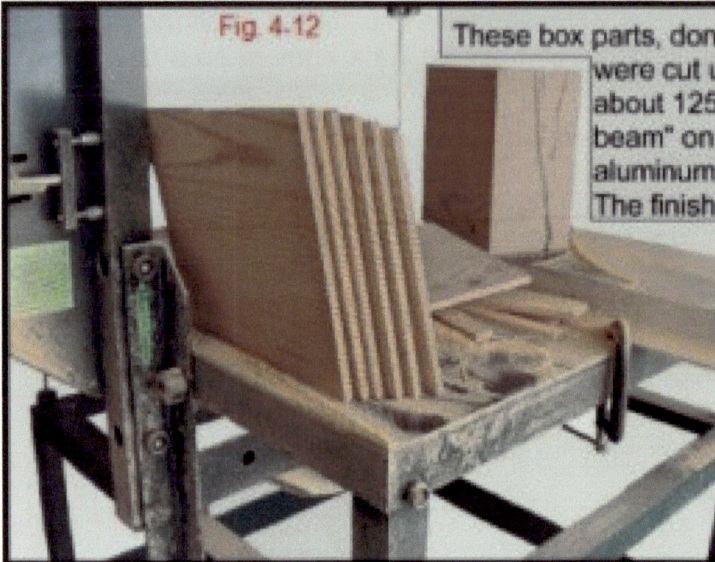

Fig. 4-12

These box parts, done with an earlier model of Corbel King, were cut using a Timberwolf 1/2" - 2TPI blade with about 125 pounds of spring tension from the "mother beam" on the far corner. The fence was a piece of aluminum angle clamped, and parallel, to the table. The finish wasn't as smooth as some prefer; nor was it cut as fast as some prefer; but at 16" wide the six sides are perfectly flat, straight, and equal in thickness. There's no reason you can't do the same with your band saw given the appropriate choice of blade and a little patience. Bear in mind that this is fundamentally a portable bandsaw weighing 65 pounds: with a lightweight table stuck on it. Your saw is probably a little beefier.

Now that your wheels are spherically crowned and properly tracking on an adequate frame with sufficient spring tension to run the blade of appropriate geometry to your project, it's time to think about how you're going to guide said workpiece through said dream-saw to get the desired result. And ain't that a mouth-full of variables! It's really hard to generalize when it comes to fences because every job has its own very specific parameters in regard to what you've got to work with, and what you want to do with it. The terms "bowing" and "cupping" are constantly used in such discussions and since they seem so often to be used interchangeably I find it necessary to illustrate **my** understanding of their meaning:

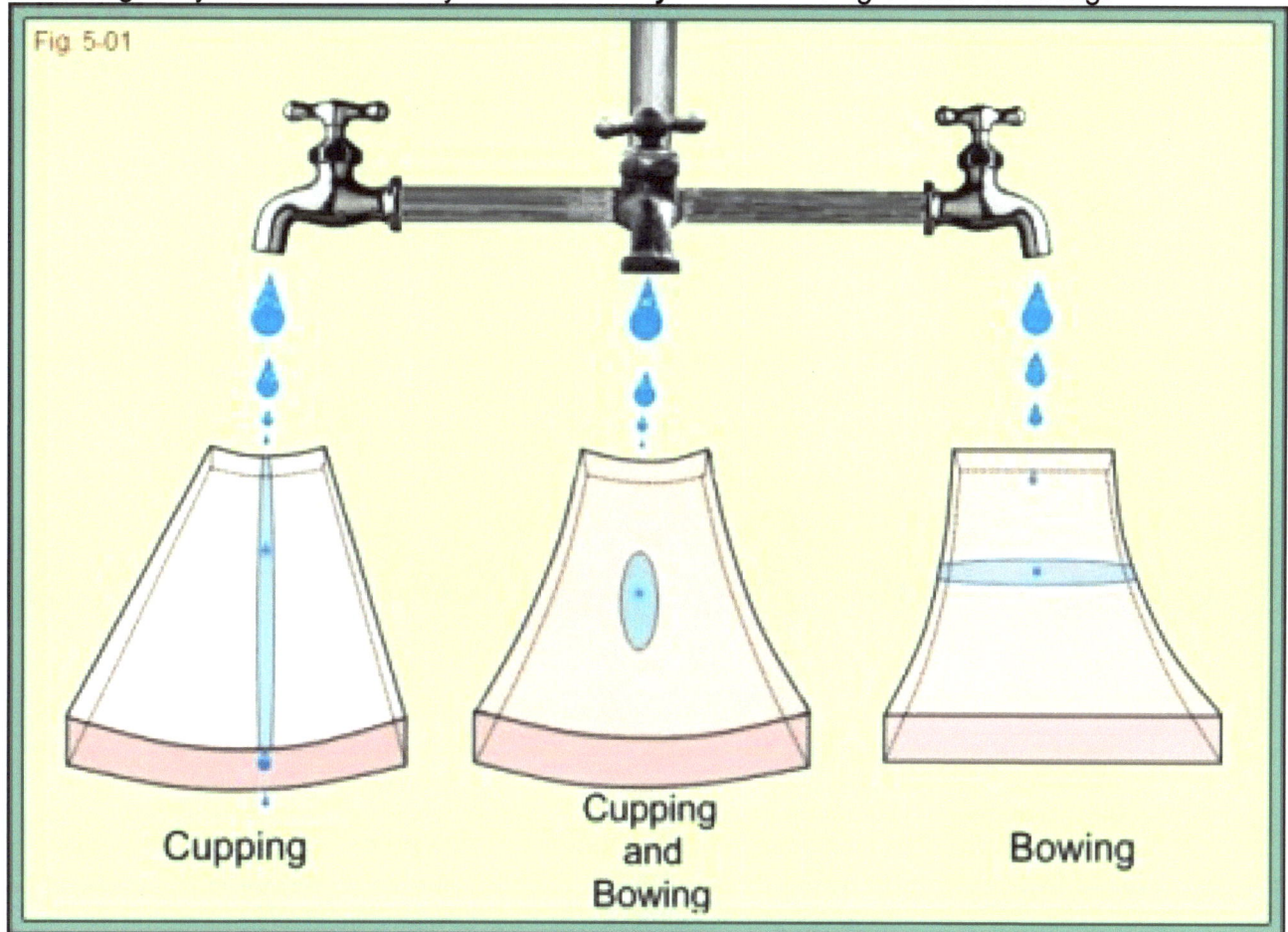

Fig 5-01

Cupping

Cupping
and
Bowing

Bowing

I don't know why, but the prettier the wood - the more it wants to cup and bow when you slice it up in a band saw. I'm not talking about blade-deflection caused cupping here, as in not-enough-tooth-set; I'm talking about wood movement as a result of liberated stresses inside the wood itself. When this form of warpage occurs **as** you're cutting it presents unique challenges to the woodworker (that being you) to keep said warpage contained until the flat, wide board of your desire is obtained. This is why we see such a diversity in fences, sleds, feed rails, trolleys, conveyors, and guide posts. Contour cuts are relatively easy compared to the fence set-ups required to make *controlled, flat, predictable,* and *straight* rip cuts. Now that home-building resaw fences has become the nation's fastest growing hobby it might be advantageous to discuss the matter like the adults we truly are, inside.

For starters; before you can effectively guide a workpiece along a fence you have to have a flat surface on said workpiece. In order to get a perpendicular edge on a workpiece you need two flat surfaces. When you're starting out with nothing more than a round log it means you have to cradle the log somehow in a sled or rail jig to define a straight, flat starting plane. There are probably millions of such home-made wooden sleds in existence today that have seen temporary use for a specific project and then set aside to gather dust; I've built a few myself. It's always a miracle if they work as planned the first time but after several passes we somehow get a flat side to work with. The only way I've found yet to do it right the first time, every time is a sliding-rail transport system that is re-usable, adaptable, and simple enough that it can be fabricated in the average woodworker's shop at a minimum of cost. Most of you already have the tools to do it and in Chapter Seven you'll find drawings and instructions.

Fig. 5-02

Although the scale is much smaller, it works like a mill and we couldn't find any degree of error in the finished cut. This was done with the second prototype of Corbel King so please excuse the funky five-piece front safety cover; the rail worked nicely. As you can see, it has leg attachments that clamp onto the ends of the rail to support some pretty heavy logs. The whole saw, table and all, only weighed about 100 pounds and some of the logs we ran with it weighed over 300 pounds so end support was essential. In the picture shown we didn't really need the leg attachments. The inner slide is nine feet long and the outer rail, is ten feet long. The slide is made from 2" x 6" x 2" channel and is laminated with a matte-finish Wilsonart for reduced friction. The rail assembly is attached to the saw by means of two 2-way clamp brackets that grab both the saw's table-top and the rail at whatever location you prefer. The log is trapped and held in the slide by means of a three-point screw arrangement that bites into the ends of the log and braces itself inside the slide's legs adjustably with sharp-tipped, pointy little 1/4 - 20 screws on a knob.

Fig 5-03

If your workpiece is less than 6" in diameter it's light enough to run the 6" channel vertically along a standard (high) fence, using the same clamp system; or you could vertically clamp the whole slide-rail assembly to the table. It makes its own fence.

Fig 5-04

The best thing about this rail system is that you can make your second cut perpendicular to the first by simply laying the first flat cut straddling the top of the rail and cut off the over-hanging log on a pre-determined line. After that it's all fence-guided truing and slicing. For truing, the only thing you need is a straight piece of anything that can clamp to a table; but slicing, as in veneer, requires a bit more precision and control. Big chunks of timber will lay relatively flat on the table and if your blade is perpendicular thereto you'll get consistent thicknesses but as the timber starts to look more like a wide board you need a fence almost as wide as the board you're cutting to insure the planes of each remain parallel throughout the cut.

Fig. 5-05

No matter how much tension you apply to a band saw blade it is still possible to overfeed it and cause the blade to twist at the beginning of a cut. The wider the blade and the more you tension it, the harder it is to do; but it's all relative and it's still quite possible to overpower any combination of tension/blade-width with sufficient muscling of the initial entry cut. Narrow kerf blades will better support themselves in a straight line once the blade is embedded in the kerf, but until then it's best to take it slow. You'll soon enough get a feel for how much feed pressure the cut can sustain. In this respect, once again, you're better off with narrower, wide-set blades because they're more inclined to straighten themselves out after a hasty start. Blades of low set angle don't allow that luxury.

While we're on the subject; remember that a smooth-cutting, narrow-set blade IS a fence, and will guide the workpiece in a straight line; so once you start a cut with an un-set blade it will continue in that direction, regardless of where the fence is pointed. If the fence interferes with the path of the blade the machine will explode in a cloud of dust and gore. If the fence veers away from the blade it's OK and you can always use an extra door-wedge. It is therefore advisable to look closely at the direction your blade is pointed during the first blade-width of a cut and see that it is parallel. As in the art of fishing: presentation is everything. Until you get a "feel" for the maximum feed rate your operation can sustain without deflection it is best to go slow, especially during the initial 1/2" or so when the blade is so free to twist in the sullen breeze.

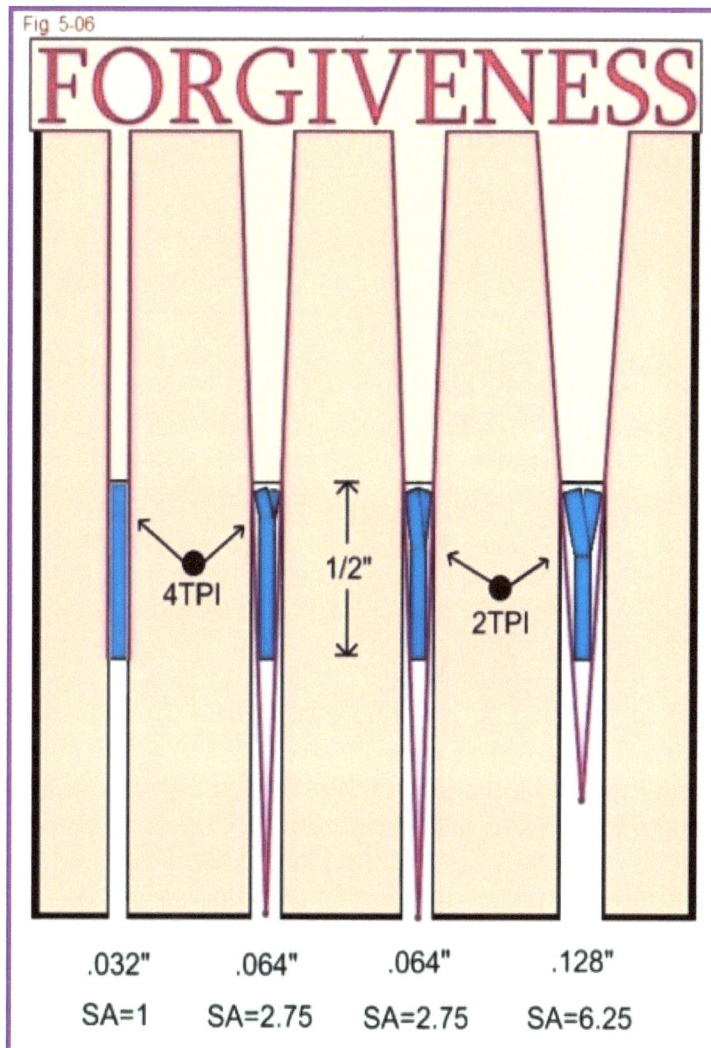

Fig 5-06

FORGIVENESS

4TPI
1/2"
2TPI

.032"   .064"   .064"   .128"

SA=1   SA=2.75   SA=2.75   SA=6.25

Think of the **blade** as a fence. Un-set; it's a straight, rigid fence. Set; it's a variable fence, and the more set, the more variable. The more set your blade has, the more forgiveness it will have for a mis-aligned fence; and the rougher the finished cut will be. You have to decide whether you want to spend your time planing and sanding or aligning your fence to perfection. It's all the same to me.

Smoothness of cut is not determined by the number of teeth per inch but by the amount of set and the shape of the tooth. Carbide-tipped blades are obviously smoother cutting because the teeth are squared and don't have that sharp little triangle sticking out the side. Blades 2 and 3 from the above illustration will leave pretty much the same rough finish despite the fact that #3 has a finer pitch (you have to look close to see it). It's the set angle that determines how rough a finish is likely to be, not the number of TPI, as some would have you believe. The "band saw lines" are as much a result of the blade constantly changing directions to compensate for deflectionary forces as they are from non-responsive tensioner flappage. Accelerating tension-spring response time doesn't completely eliminate the cut's roughness altogether but it definitely reduces the depth of its scratches.

Assuming you now have the blade, frame, tension, and power to make straight flat cuts the only thing stopping you from making perfectly flat consistent veneer is a fence that can line up perfectly with the plane of your blade's kerf. And that's easy. Just take some 2" x 4" aluminum tubing, a little 1/2" x 3/4" bar stock, a few screws, some 2" x 2" x 1/8" angle, and make one like this:

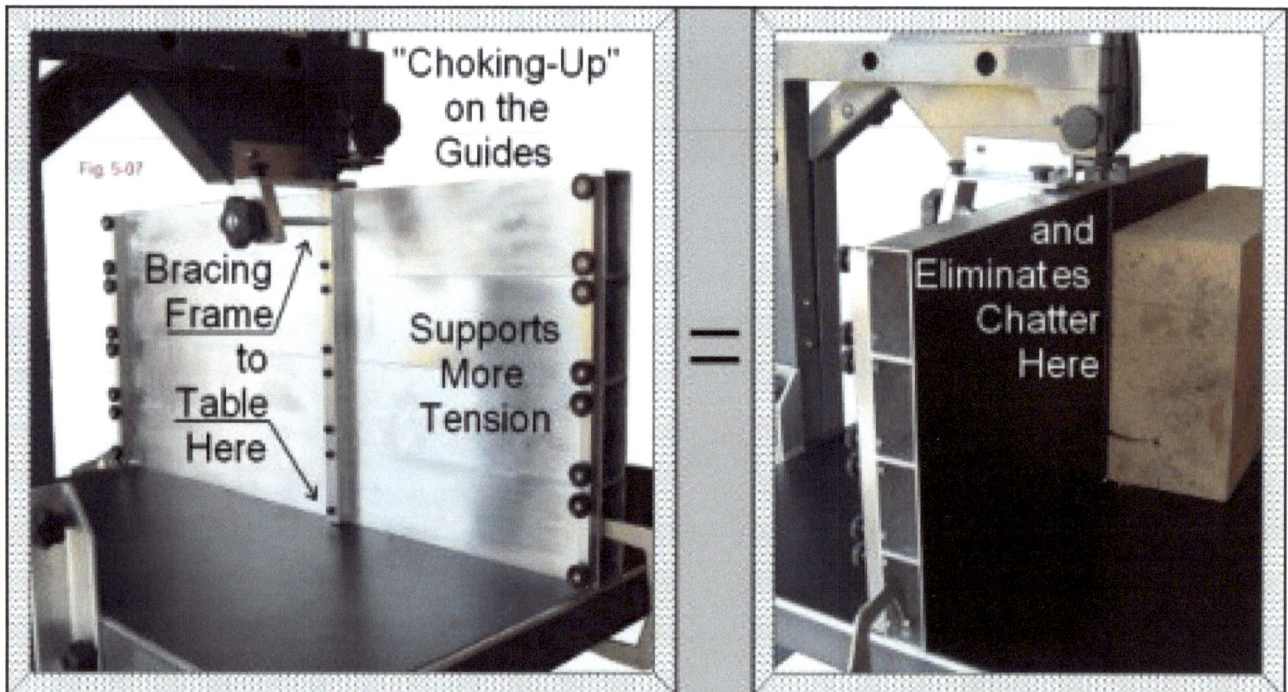

This was my first attempt and it worked nicely to slice veneer off large timbers but it was limited in function by the short slide adjustment on the top support rail which only accommodated rip cuts up to 3" thick. The welding clamps worked nicely to align the horizontal axis; I simply squared it to the table-top at the desired veneer thickness and clamped them down. It still worked for wider cuts but beyond 3" the top support had to be disconnected and one was counting on the workpiece to be perpendicular. In retrospect: the laminated surface wasn't really necessary; the vertical adjustment screw *pushed*, but couldn't *pull* the top of the fence into vertical alignment which made it tricky to set. Its major faults were that: (1) it was overly elaborate, costly, and time-consuming to build: (2) if the ripped veneer beyond the blade bowed or cupped it pushed the workpiece off the fence: and (3) even when the cut was perfect, the veneer piece would taper off at the end as the workpiece closed into its new kerf, tipping the trailing end away from the fence.

Obviously this can be avoided by holding the *mother-piece* closely against the fence; but when you get towards the end of the cut there is no longer a flat surface to guide you, and invariably the trailing edge will taper off as shown below. In some cases the *push-off* phenomenon is too strong or the operator's attention strays for a moment and blade is allowed to stray even sooner. In the first case the majority of the veneer is still usable if one cuts a little off the end; but in the latter case the veneer is ruined and the operator must start over, wasting good wood.

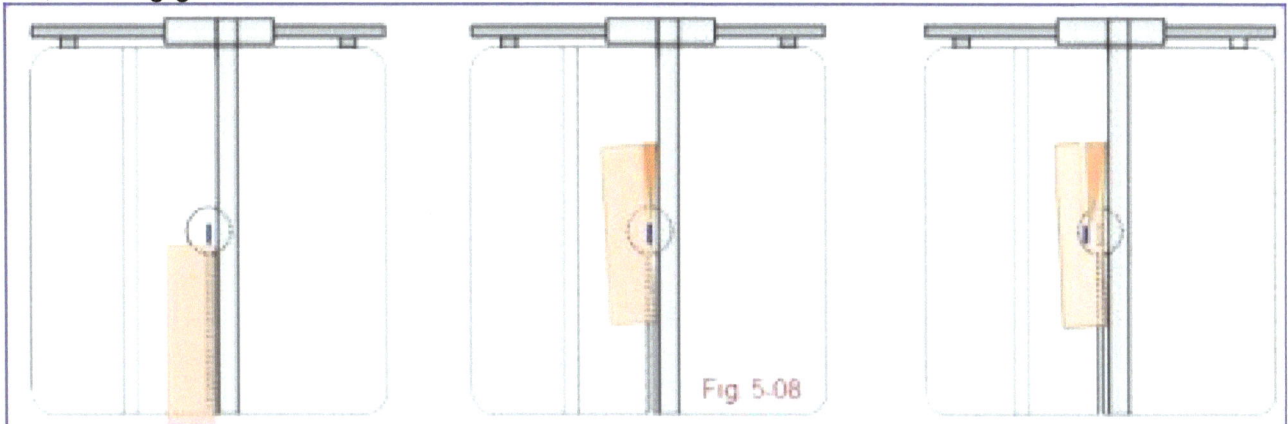

Fig 5-08

There's another phenomenon that can occur when one liberates the internal stresses inherent in figured hardwood and you'll see it when you're about half way through a rip cut. The off-cut veneer and sometimes even the motherboard will warp making it nearly impossible to follow a full-length fence. If your blade has enough set it might cut the subsequent bow; but if not, the blade will follow the motherboard and you've got yourself a real mess as seen above. The drawing above is slightly exaggerated, but not by much. In this case, even a well set blade won't save the cut from being a disaster and the only solution is to cut the fence off short at a point right behind the blade, so the split pieces can go their separate ways unimpeded. All-in-all I think half-fences are preferable to full-length fences under any circumstances and I don't understand the popularity of the common bandsaw fence at all. I especially can't understand any fence that isn't clamped to the table at both ends because surely the operator is pushing his workpiece into the fence with varying degrees of aggressiveness that could result in **fence** deflection.

It's not uncommon for woodworkers to use a shortened fence for such operations and eliminate push-off by stopping the fence immediately after the blade. It certainly solves the push-off problem and constitutes a major improvement over full-length fences for veneer slicing but still leaves one with the end-cut-tapering problem, where you totally lose fence guidance at the end of the cut and have to free-hand guide the last couple inches. We all know what happens when we're left to eye-ball woodwork; don't we. It's never quite perfect. I'm told it's not a problem for professional woodworkers who have a 12" planer to straighten it out, but what about guys who want to rip more than 12" wide boards or guys who just don't want to re-set their planer between rip cuts?

Fence deflection can be avoided by pre-determining the blade's orientation and aligning the fence to that (the right way) or doing away with the fence altogether and substituting it with a post vertically aligned to the blade and supported, top and bottom, by the table-top and the blade guide's extension post. While the vertical post solution is a convenient short cut for re-sawing thin boards free-hand, the resulting horizontal plane isn't much better than you'd get hand-feeding a scribed straight line and should only be viewed as an emergency measure when quality is not an issue.

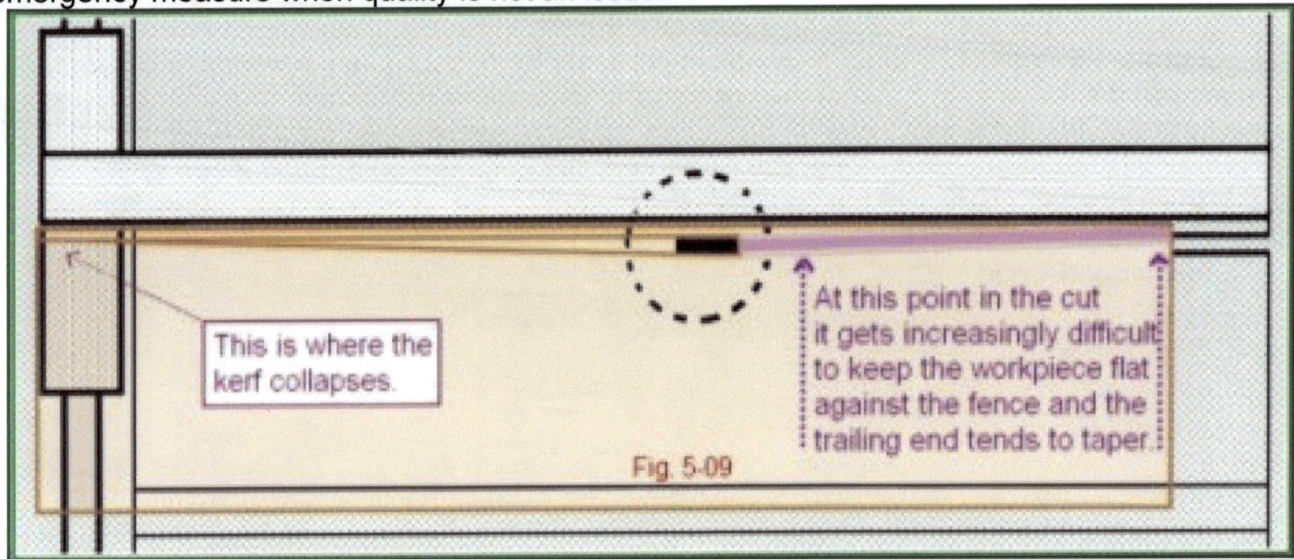

This is where the kerf collapses.

At this point in the cut it gets increasingly difficult to keep the workpiece flat against the fence and the trailing end tends to taper.

Fig. 5-09

While we're on the subject of things I don't understand; why, when the bandsaw's chief advantage over table saws' ripping ability is its cut depth, do they equip them with 3" and 4" fences? I mean, the fence is only there for rip cuts, right? Why would anyone want to rip a 4" cut on a bandsaw when the table saw can do it so much faster and smoother? I'm sure many of you must have asked that question while you were building your home-made plywood resaw fences. Why would anyone build a bandsaw with 12" of re-saw capacity and equip it with a 4" fence? My only guess is that nobody's thought to make a modular fence that stacks fence support incrementally.

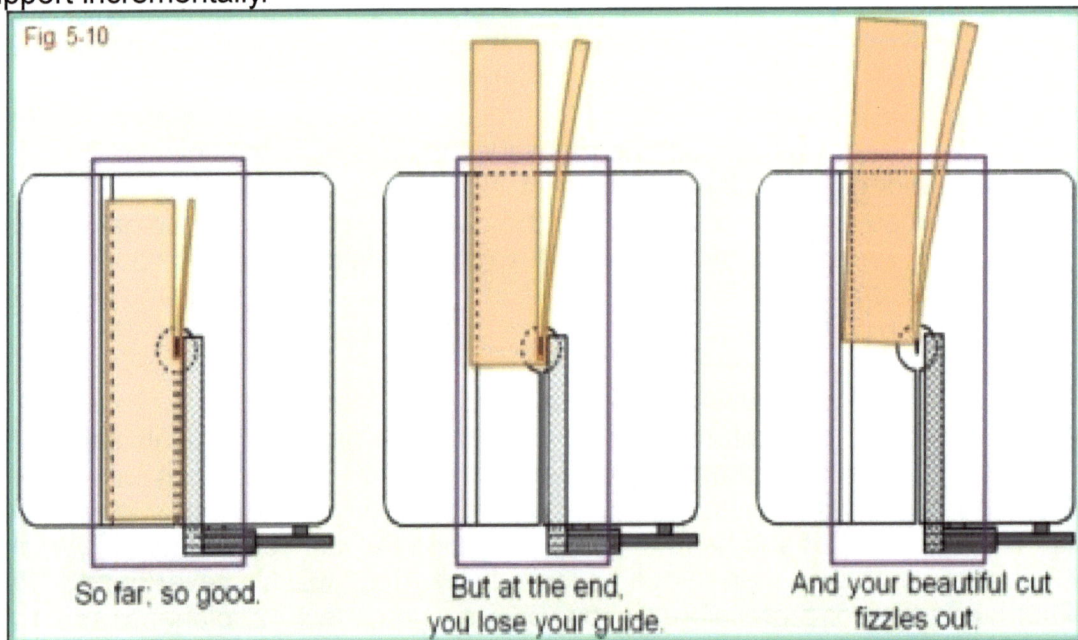

Fig 5-10

So far; so good.

But at the end, you lose your guide.

And your beautiful cut fizzles out.

This cutting-planing-cutting-planing sequence seems to be the accepted way of doing repetitive veneer cuts but I seriously question the labor-and-material economy of this procedure. I can understand why one might want to plane a piece to have it ready for glue-up, but not everybody has a 12" planer and it wouldn't help at all if you're cutting 16" wide veneers anyway. Would it not be better to make successive cuts, one after the other, that were already totally flat?

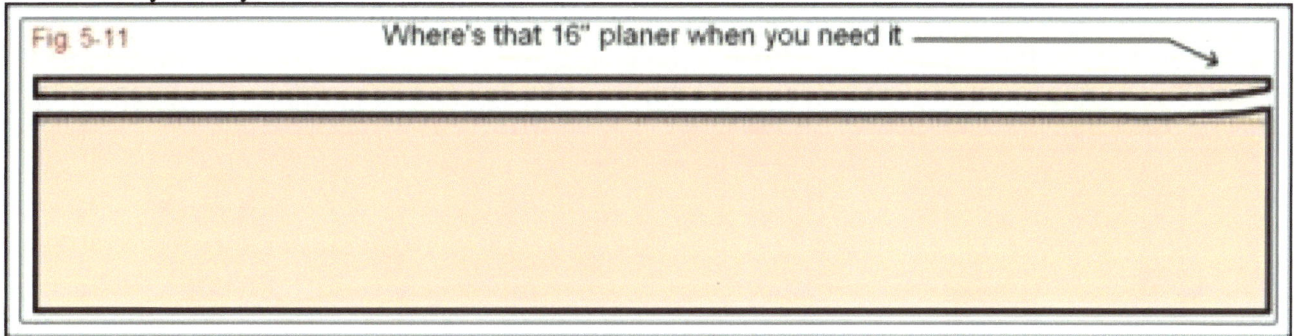

Fig. 5-11   Where's that 16" planer when you need it ⟶

When you have nice, dry wood that doesn't curl up after it's cut, you can use a full-length fence to cut perfect 15" wide veneers like this:

Fig. 5-12

But when you're cutting thin veneer with a vicious curl you have to go a step further by fence-guiding the workpiece all the way through the cut. That's where the *vestigial* fence comes in. The vestigial fence, or stub fence, or kerf-chaser (I haven't decided yet) is located immediately behind the blade and follows the plane of the **blade,** not the plane of the **fence**. It prevents the mother board from wandering at the end of a cut and allows the off-cut veneer to **ooze** out the other side and fall flat on the table.

Fig 5-13

The stub fence on SliceMiester I isn't really pink. I highlighted the photo so it would be easier to see. The picture on the right with my smiley face on it, looking very professional in my white lab coat, was taken as I cut some primo cherry 15" wide into 1/8" veneer. You can see the vicious cupping taking place before I was 6" into the cut. I use awls instead of fingers when I'm working so close to the blade for the obvious reason that awls are easier to replace than fingers if the blade wanders outside its intended path. What I really wanted at the time was an auto-feeder to hold the blade consistently and aggressively flat against the fence while feeding forward at the same time. We're working on that and you'll see it on future embodiments of the SliceMiester feeder as well as Chapter Seven in this book.

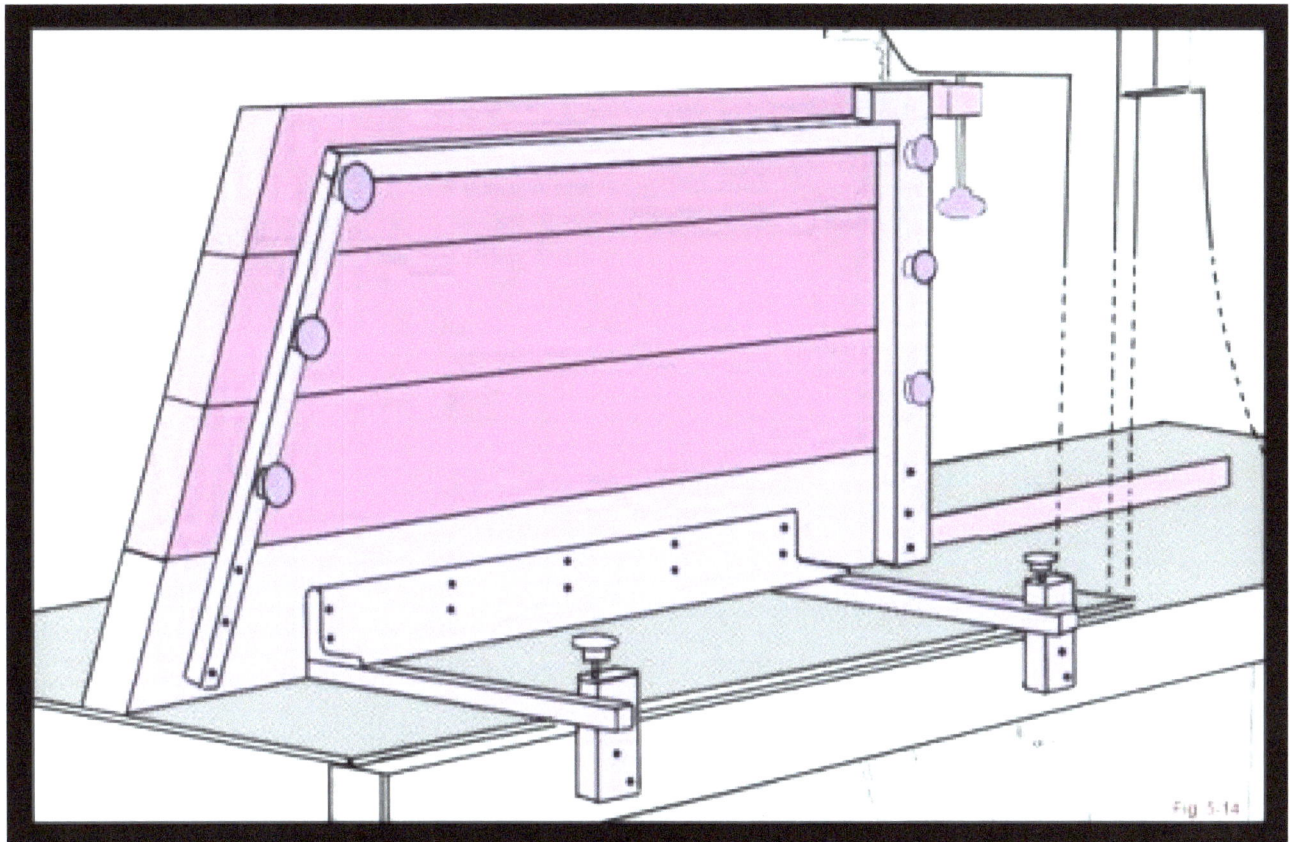

Fig 5-14

The drawing above, showing the backside of a SliceMiester's fence, is highlighted to distinguish between the basic bracket and its add-on modules. This was an all-purpose modular fence that: (1) accommodated the upper blade guide in 4" increments; (2) snapped out in seconds by means of the three vertically-oriented knurled knobs shown; (3) facilitated x-y adjustability in infinite increments; and (4) re-enforced the frame/table relationship for greater blade tension and the elimination of "chatter". When changing blades, the veneer thickness setting could be re-calibrated with 1/2" x 3/4" bars that clamp to the table. Returning to a previous vertical setting was easy to see by the indentations on the top-angle/rail and a scribed index on the base extensions. Resetting to another thickness (up to 8") was as simple as putting it where you wanted it and clamping it down.

The first SliceMiester prototype had a three-part table that adjusted to position the stub fence but you can also do it on your cast iron table by milling a slot behind, and in line with, the blade. I used stainless steel for its hardness and cut the piece shown with a Dremel, using a cut-off wheel. It's not recommended but it's all I had to work with at the time; you'd do better to have it machined by the same guy who cuts your slot. The leading two inches of the stub fence are free from the top and it can be adjusted, by bending, to capture the kerf precisely. The rest of the fence is subject to modification and improvement, depending on your tools and skill level. You could also over-lay an auxiliary fence jig on your existing table to make it more adaptable to modular and stub fences and I'll describe that in greater detail also when we get to Chapter Seven.

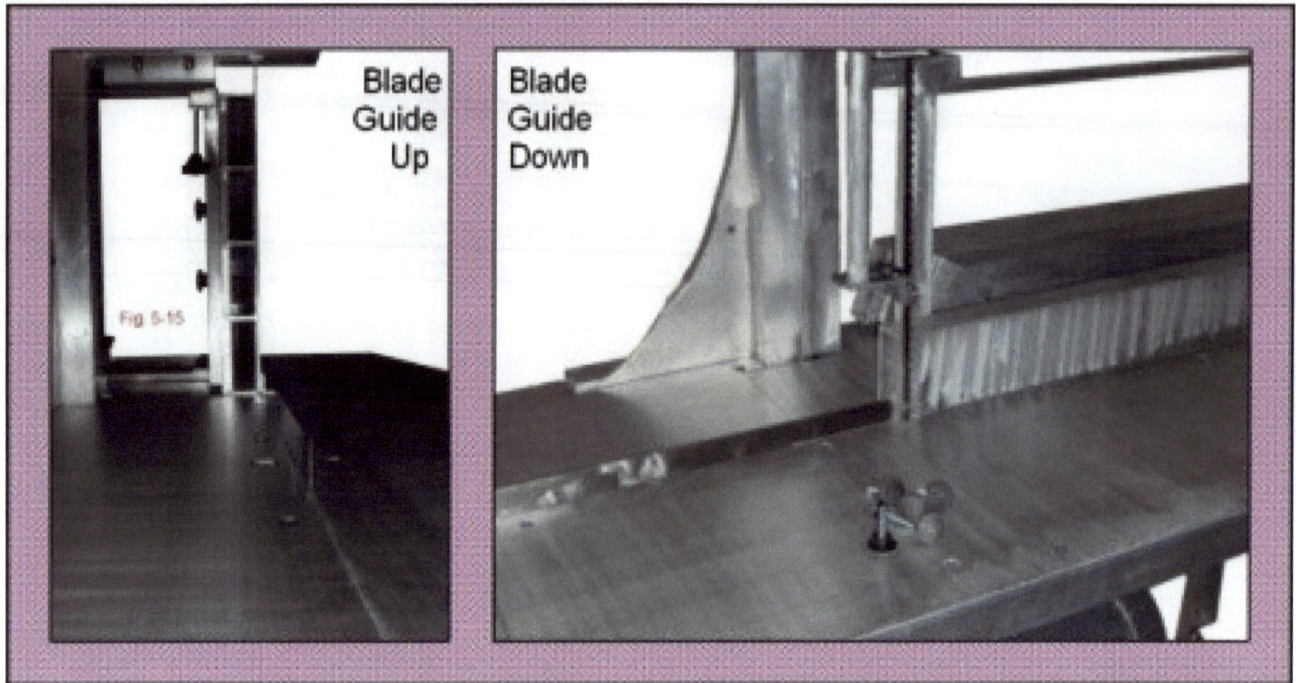

Blade Guide Up

Blade Guide Down

Fig 5-15

I don't know if it's obvious or not, but the basic reasoning behind the use of stacked tubing between the vertical supports and the blade is that the inside blade guide roller has to have some space between the blade and the fence. You can't have a fence any higher than the upper blade guide when you're cutting less than the diameter of that top inside roller. Removing a section of tubing lets you bring the upper blade guide down 4" , and closer to your workpiece.

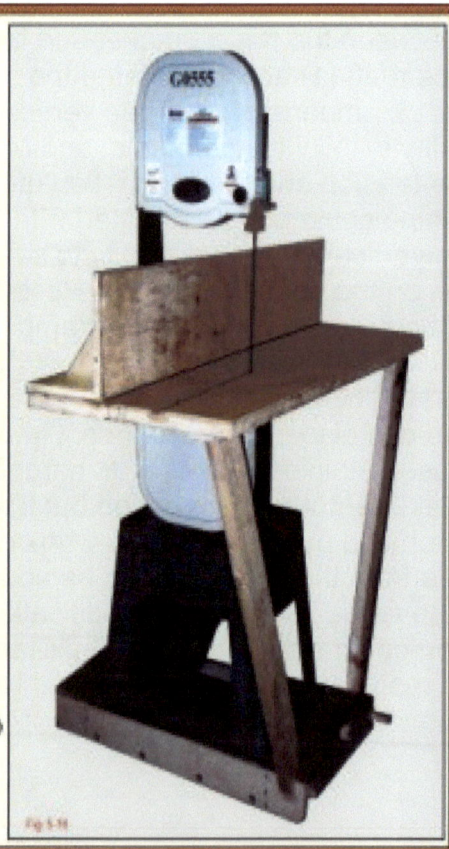

How many hundreds of thousands of home-made table extensions and high re-saw fences do you suppose are built every year in just the United States alone? While there may not be many as well thought-out as this one by Mark Rios, they all attempt to extend the ridiculously tiny (but tiltable) machined cast iron OEM table and add some height to the fence. Mark even went a step further by incorporating his mobile base into the over-all upgrade. Naturally, I question his choice of material to execute the upgrade, but there's no denying the design-genuity that went into it. One would think such an obvious improvement would be available as an option from the factory or as an after-market product with universally adaptable mountings. But no, it's something we're going to have to do on our own until someday an unemployed shade-tree entrepreneur comes along with a jig so slick and universal that it becomes a marketable product. Don't look to the saws' manufacturer to lead the way in this innovative surge. They regard such things as unprofitable niche markets and won't help so long as you continue to buy their saws, as is. Thar's gold in them thar hills, fellas, go get it. Just do it in aluminum next time.

CHAPTER SIX
Proportion is all relative to scale.

      Now that we see how inter-related the individual components of a band saw are, we begin to see how a band saw's performance is only as strong as its weakest link. I think it's pretty safe to say that most instances of mis-proportioning are more the result of economic considerations than engineering. It's not hard to see why a company that makes band saws in five different sizes, ranked by wheel diameter, would want to use the same blade guides for all of them rather than tool up to make them in five different sizes. In most cases the blade guides are over-built for all of them anyway and the savings associated with a one-size-fits-all strategy goes un-noticed by the consumer. Where the one-size-fits-all design strategy causes confusion and misconceptions for the consumer is when it is applied to size-critical components like the tension spring, its bracket, slide, and adjusting screw. Trunnions, table dimensions, and frame elements are often shared by larger and smaller saws resulting in 14" saws having the heart of a 20" saw or 24" saws having the heart of a 10" saw.

      The matter of motor size is probably the most confusing of all in this respect. Disregarding sawmills and other industrial applications, anything from 1/3HP to 3HP will turn the wheels on a consumer-sized bandsaw and it's purely a matter of individual discretion what size suits your needs. Three HP, triple phase motors are great if you can justify the weight, expense, power consumption, and wiring requirements but unless your bandsaw has the requisite blade, blade transport, material transport (fork lift), and cut depth to actually **use** that kind of power, it's a waste of resources. In most cases there just isn't enough need-for-speed to justify the expense of such high-speed saws other than the bragging rights of being faster than the shop next door. Which is OK, too, if that turns you on. It's purely discretionary. Check out my jet-powered saw. It'll beat your saw any day. And I've got the pink slip.

      I started out using 1/2HP ODP motors on my portables but occasionally got complaints that the motor was bogging down and upped it to a 3/4HP ODP. When I got too many calls

Fig. 6-01

about the motor not turning on I started using 3/4HP TEFC motors and I still got calls complaining that the $14 Delta switches were not working so I started using $5 cheapo rocker switches from the local hardware store and haven't had a complaint since. What's of interest here is that I had a couple, and that's not many, guys express a desire for a little more power than the ¾ TEFC Dayton motors because they had bogged down a couple times. One of them went so far as to install a 1HP on his own and reported back that he didn't mind the added weight but surely appreciated the extra oomph when doing a production run of arches so I started using 1HP on all my Titan saws and gotten no complaints about the extra weight. The point is; there's no magic *right* size but the one that's proportional to the job you put it to. On a stationary saw that you're not going to carry around all day a little extra HP won't hurt. It won't compensate for any other weaknesses in the chain, however.

Speaking of chains; there's a growing segment of the logging and milling community who think chain-saws are wayyy under-powered and have the shown enough initiative to actually do something about it. They have found, as I did, that the main-stream manufacturers of woodworking machinery are totally unresponsive to their consumers' demand for higher performance. There seems to be two divergent design philosophies at work here where one is dominated by children making tools for children and the other consists of adults making toys for adults. On the fence between them are children making tools for adults and adults making toys for children. If you can determine which is which you should seriously consider designing your own tools for woodworking, because you have a definite "edge" on the rest of us.

As difficult as it is to work with wood, you'd think we'd see far more innovation going into the design of woodworking tools. I can't think of another medium that presents the design challenges facing woodworkers every time they endeavor to shape and join a material with so many idiosyncrasies as wood. Like no others, woodworkers deal with a medium that substantially changes in dimensions and shape with every change in temperature or humidity. Yet, with all their superior fabrication skills, woodworkers seem reluctant to deal with any kind of metal work; even when metal is obviously called for in the choice of jig-making media. At the risk of over-generalizing on this admittedly subjective analysis of such a diverse demographic, I think there's an innovation gap between garage-bound woodworkers and the wider "tradesmen" demographics of the construction trades.

As you can well imagine, the guys in the following picture, who manufacture chopping blocks for a living, don't have time to stand around waiting for some wimpy foreign-made chain saw to slowly work its way through a little 48" tree stump. Taiwan won't build specialized machines like this so you have to do it yourself, like these guys did. I don't see any signs of nitro fuel lying around or being used so I assume this is a local chopping block distributor. I have no doubt that someday we'll see hemi-powered portable bandsaws that will absolutely "smoke" these guys in the races (what with the thinner kerf and all). No flannel shirts or suspenders here; just lean, mean, manufacturing efficiency. "Gettin' 'er done!" the American way. One has to ask the inevitable question doesn't one? Were they auto mechanics first or stump-cutters? I submit they're Americans performing their pre-ordained national duty; to innovate. Making your own tools is a tradition in the "New World" that goes back further than beer; and I fear we've forgotten that lately. I don't know who these guys are, how the saws were made, nor whether they hand-forged and machined their custom parts to connect a chain bar to the drive shaft of a V8; it's irrelevant.

What is relevant is the amount of time and money tradesmen will invest in the tools of their trade. I might also add that the degree of socialization surrounding the exposition of such masterpieces of engineering are merely the tip of a business network that connects every trade in the local community. It's fun. And it's open to anyone with a desire to know what makes things tick. The point, as it relates to hot-rodding your bandsaw, is that you're not alone and you're not expected to know the intricacies of welding, machining, or metal forging to design proprietary tools for your woodworking operation. All you need is the ability to design and layout the basic framework of a new tool in another media besides wood. It's not as frustrating or difficult as you might expect and I'll discuss some basic aluminum-working techniques that you can do with the woodworking tools you already have in Chapter Seven. For now let's talk about little things you can do to beef-up the weakest links in the mechanical chain of your bandsaw.

It's not hard to see that the weakest link here is the operator. In every picture I've seen of guys running V8 chain saws, the guys were as big and beefy as they come; yet they still had to struggle to get the saw to the job. The only possible upgrade, therefore , in this situation would be to replace the two operators with one well-trained 800 pound gorilla. That ought to do it.

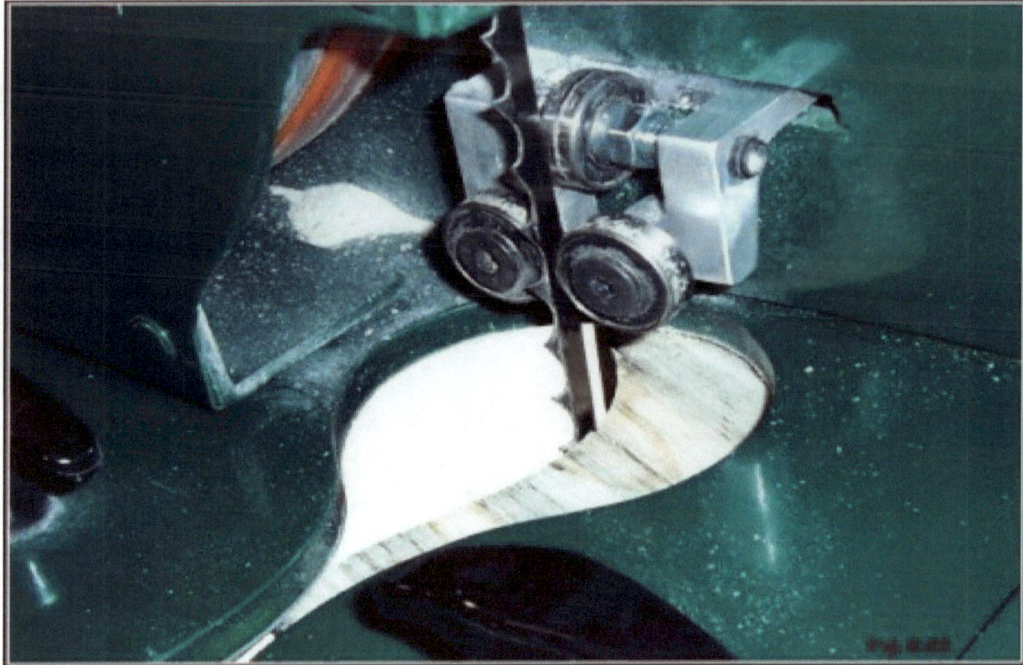

For bandsaws, too, it's all about the length of time it takes to make a cut. Cut time is what determines the proportions to which your bandsaw should be designed. You can cut just as deep with a 1/2" blade as you can with a 2" blade; if you're not in a hurry. If you want to hurry you'll need a wider blade. If you want to run wide blades you'll need more tension. If you want more tension you'll need a stronger spring. If you want a stronger spring you'll need a beefier tensioning bracket. And so forth and so on. This chapter is about where and how you can push your saw's "envelope" proportionally.

I'm guessing you want the most versatility you can get for the least amount of money; and you'd be right in thinking so. Assuming we get what we pay for; the more a saw costs - the more versatility we should expect to get out of it; but that's not how it's working, I'm sad to say. Ignoring their consumers' demands for true versatility in bandsaw design, the leading brands focus their design efforts on the manufacture of strong-but-cheap, compact frames with less versatility. In the ten years that I've been in this business I've learned that bandsaws comprise a long series of compromises and that maximum versatility starts with a big frame around a lot of throat area. A fast rate of cut requires a strong frame. At present our only options for big **and** strong are cost prohibitive for most consumers (see Tannewitz 36") and the giants of the bandsaw industry have yet to adopt the necessary three-wheel design technology to make that quantum leap. Moreover; big and strong, in regard to band saw frames, don't necessarily have to go together. A **quality** bandsaw would be cost-effective and specifically designed to meet specific goals. What do you expect the saw to do? We all have have specific goals. By mass-producing clones of the same basic bandsaw design the tool industry has left the consumer on his own to improvise sleds, fences, just about everything but the frame.

It's common practice to rank bandsaws according to their wheel diameter; 12", 14", 18", 36", etc. Wheel size is certainly a significant factor in the classification of bandsaws but by no means the only factor. Nominal ranking characteristics should go on to include cut depth, throat width, and blade width capacity. While the cheapest of plastic-wheeled bench-top saws will run a 1/8" blade, it takes considerably more beef in every component to adequately support the operation of 1" wide band saw blades. When you look at the frames found on most commercially available bandsaws it's easy to imagine them ripping big timbers with a 1" blade but you'd most likely be wrong and find that the spring won't support that level of tension without red-lining. Manufacturers don't like to confuse their customers with too much information. They don't advertise the spring rates of their saws' tensioners; but the standard for 14" saws, being about 2.5" or 3" long, 5/8" in diameter on the outside with an ID of roughly 3/8", range anywhere from 100 pounds/inch to 500 pounds/inch. Bigger resaws can run up close to 2,000 pounds/inch. Absolute poundage is somewhat less because most tensioning springs don't have a full inch of free play however; but still - that's quite a range.

Another area of general weakness I see in most band saws is in the blade guides. There is a pervasive lack of regard for friction in the majority of them that indicates their designers have no sense of mechanical efficiency and energy conservation. I'll never understand why thrust support bearings are commonly mounted broadside to the blade; is there any advantage to rubbing the back of the blade? Are they deliberately introducing "spin" for any reason? Surely they don't think widening the area of contact with the blade somehow "supports" it any better. I think their designers' idea was to prolong the life of the bearings by reducing the rate of rotation and letting the blade take the wear and tear; but radial bearings only cost $5, so it doesn't make sense. I know, it's a small thing, but if you can find a way to get your thrust support bearing's plane to rotate in line with the plane of your blade it would give you a small boost in usable horsepower.

The next thing you could do to boost horsepower is to eliminate friction in the guide rollers, those four things that pinch the side of the blade while it's flying along at 76'/second. First; I should say that if you're using the guides (that's blocks, rollers, hickory sticks - whatever pinches the sides of your blade) to keep the blade running straight; you're wasting energy, creating friction/heat, and draining horsepower. The wheels are supposed to keep the blade straight; the guides are only there *to guide.* Unless you're forcing the blade to turn against its will, the guides should only be used as a visual aid, indicating the kerf's location relative to the wheels. In other words: if your workpiece is pushing the blade to left or right of its straight path between the wheels; it will be rubbing against one or the other guide, creating unnecessary friction. By referring to guides as visual aids I'm saying that if you keep the blade from touching **either** guide, you're not only conserving energy, you're insuring there isn't a bow happening down inside the kerf.

Radius turns are a different story. In this case the guides serve the critical function of forcing the blade into an un-natural twist and it is expected that great friction will result. The composition of the guides becomes very important in this situation and radial bearings become the obvious choice for reducing friction. When you push them hard enough you will feel heat building up in guide **rollers,** but it's better to heat the rollers than to heat your blade. You create less over-all friction and put less demand on your motor (again - boosting horsepower) by eliminating friction at its source.

You can probably make your own blade guides better than the ones they currently offer and they'd cost a whole lot less. It isn't nearly as hard as you'd think to fabricate parts like this and I've been doing it with a jig saw and a drill for years now. This is my idea of "close-tolerance" work so I center punch and drill pilot holes for all the tapped holes. I forget why, but the corners shown here were filed square; I normally drill a 5/16" hole to turn corners and file the straight lines to make it look like it was done by adults, leaving the radius un-touched. Square corners looked neater that day, I guess.

It's basically two forks and any configuration that fits your saw will serve to support thrust and guide the blade simultaneously. I use 2" x 6" x 2" aluminum channel to make blade guides that work on any blade from 1/4" to 3/4". The alloy they use for this extrusion has good *memory* and you can spread the forks quite a bit before they lose their ability to spring back . They give me more thousandths of an inch than I need to adjust for varying blade thicknesses (there's no more than a .020" spread from biggest to smallest). As sloppy as the production technology employed may be, the adjustments for aligning them are precise and adjustable in infinite increments to fit any blade thickness. The same design could be scaled up or down proportionally to accommodate wider or narrower blades.

The mechanical components weren't intuitive for me so I'll illustrate them here for you. After hours and hours of searching for shoulder screws with enough thread length to use in my unique applications I finally discovered the self-centering properties of the common flathead machine screw. Yes! When you bolt a radial bearing it becomes a stud-mounted bearing and turns as true, or truer, than any stud-mounted bearing. Likewise: two conical-topped lock nuts will center themselves in a radial bearing. If one were mechanically-challenged, one could even mount it broad-side to a band saw blade. The bearings I use for blade guides are: "Double-Sealed, Premium Electric Motor, Deep Groove Radial Ball Bearings; Inside Diameter: 3/8; Outside Diameter: 7/8; Width: 9/32 .MSC calls them # 35433382. (I call them BG bearings (#ET 141) but I don't retail them so it doesn't matter.)

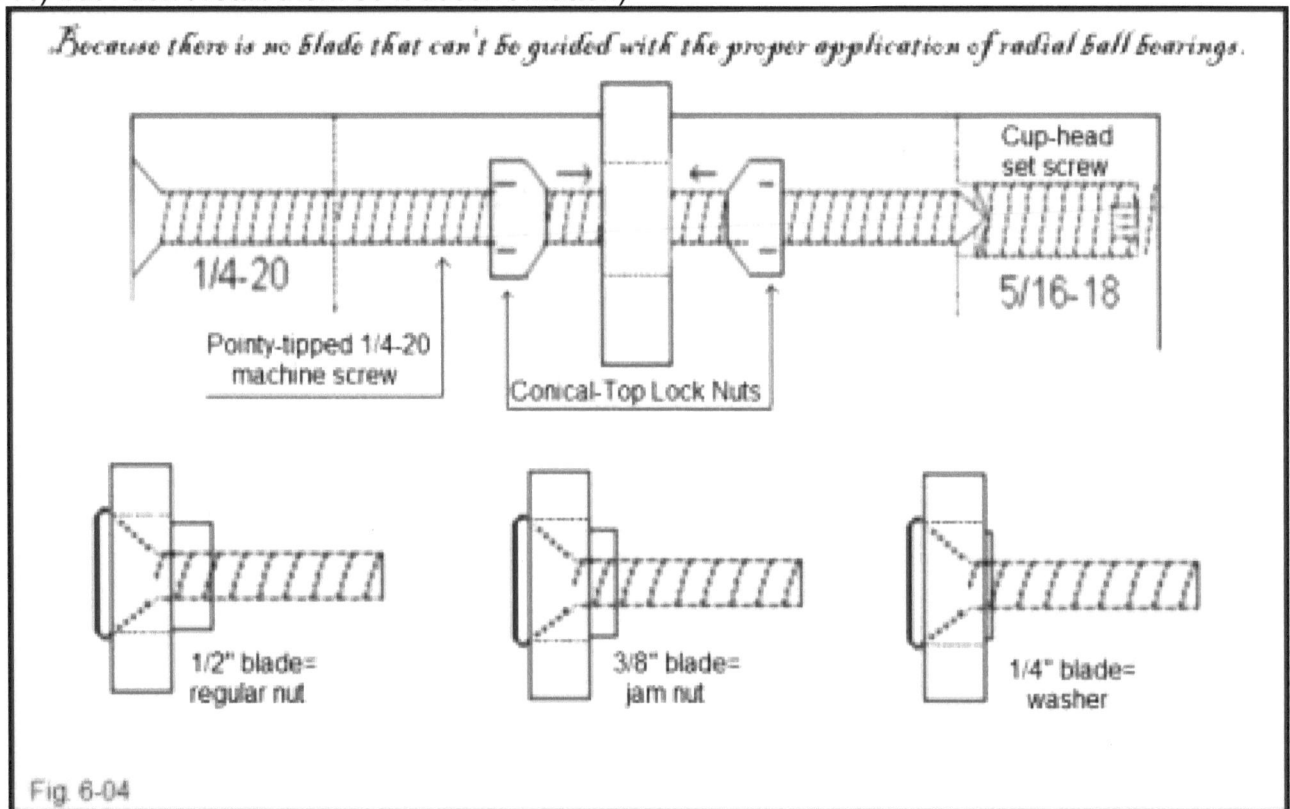

Because there is no blade that can't be guided with the proper application of radial ball bearings.

Cup-head set screw

1/4-20

Pointy-tipped 1/4-20 machine screw

Conical-Top Lock Nuts

5/16-18

1/2" blade= regular nut

3/8" blade= jam nut

1/4" blade= washer

Fig. 6-04

There are any number of ways to mount blade guides on a saw but the simplest for me is to *clamp* it down directly to the frame with a single screw. Depending on your saw's mounting to the frame you can adapt the guides with existing extensions or build your own out of round stock which is available in many diameters. The key to successful mounting is to provide for *x-y axis* alignment. It is common for most commercially available blade guide brackets to adjust forward and backward (to adapt to varying blade widths) by sliding the whole bracket on a fixed point relative to the frame. This limitation is predicated on the assumption that the wheels are fixed in their relative locations on an unadulterated frame (that is: the frame will never be bent, twisted, or mutilated (maybe-maybe not)). It also assumes the wheels' mounting holes will never wallow out and the tensioning slides won't wear (very doubtful). The vertical alignment of the blade will also vary with the amount of tension applied, and how far a reduced tension allows the blade to fly out beyond the wheels' radius. (Yes! Blades will react to centrifugal force and I let them run out a little whenever I don't need excessive tension. You can see for yourself when you set your guides for a little daylight on each side and alter the tension settings.) And, as you can see from the illustration below; when tracking is off-square, the guide bracket has to correspond (It's better to square the tracking problem, but.....) At any rate; the back of the blade guide bracket will ultimately have to adjust side-to-side for co-planar alignment with the blade. I use round stock aluminum for guide posts and attach the guides with a single bolt into the eccentrically threaded end; by which means the side-to-side location is adjusted by rotating the shaft.

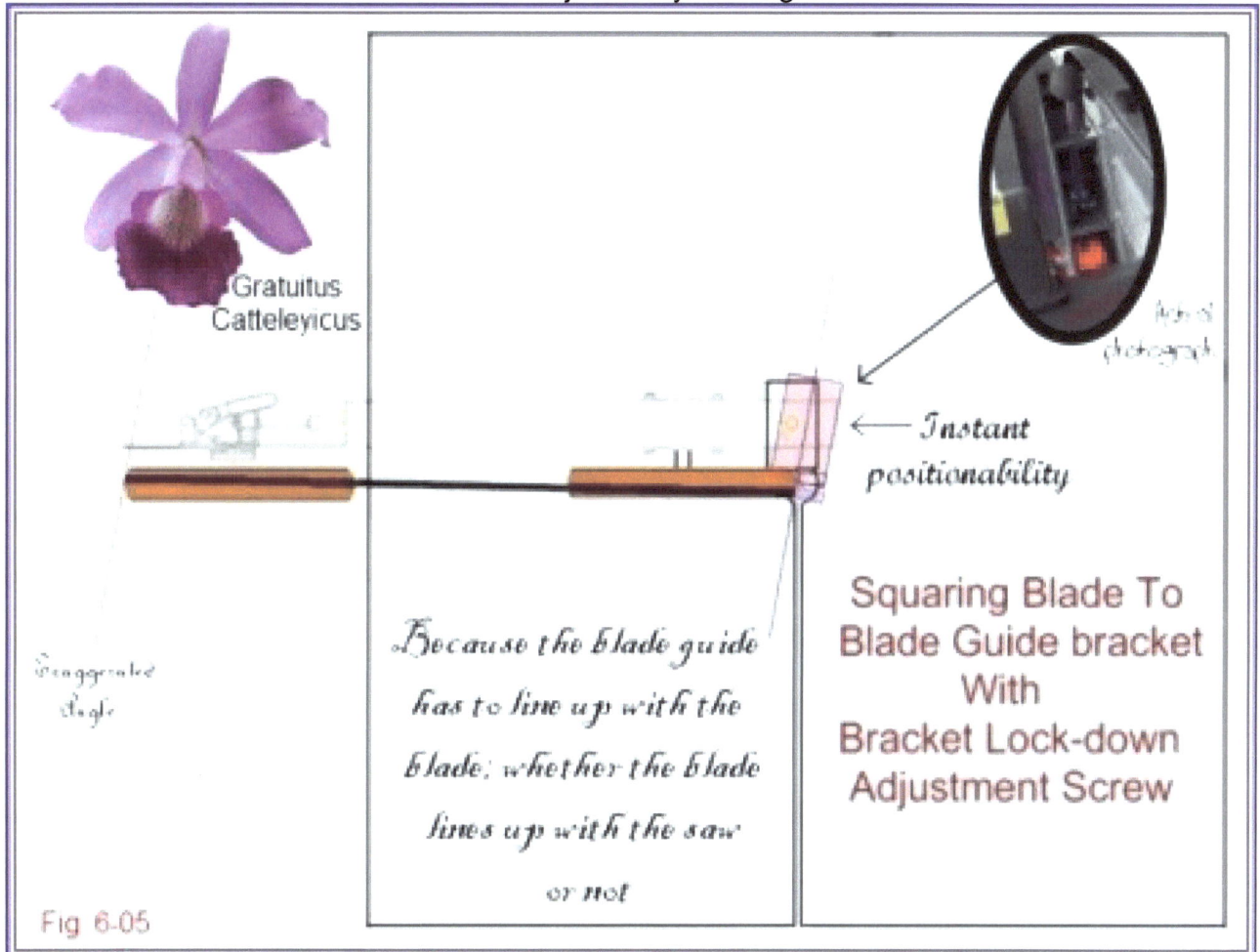

Gratuitus Catteleyicus

Because the blade guide has to line up with the blade; whether the blade lines up with the saw or not

Instant positionability

Squaring Blade To Blade Guide bracket With Bracket Lock-down Adjustment Screw

Fig 6-05

I keep using three-wheeled saws for my illustrations mainly because I have so many drawings of them. Ignore the third wheel in these instances, concentrate on the blade guides. Obi-Wan Kenobi would wave his hand and say: "You don't see the third wheel; you're interested only in the blade guide mountings." If that doesn't work, there's always the diversionary orchid. I digress.

The most egregious shortcomings that all bandsaws have in common are those itsy-bitsy little tables they come with. You have to think about that for a minute to let it set in.
Here's a saw that cuts two or three times thicker material than what a carrier-sized table saw can cut, and it comes with a 16" x 16" postage stamp table. What are they thinking? Seriously; there are a lot of possibilities for add-on after-market extensions, jigs, and

attachments. Based on my experience with the SliceMiester project I can safely say the following concept drawing would vastly improve the performance of your average band saw. The proposed dimensions are as flexible and arbitrary as the degree of complexity one might want to build into it. In its simplest form, it's a table extension; carried to its full potential, it's the basis for a very sophisticated re-saw system with multiple feed attachments for veneer cuts or bowl-blanks.

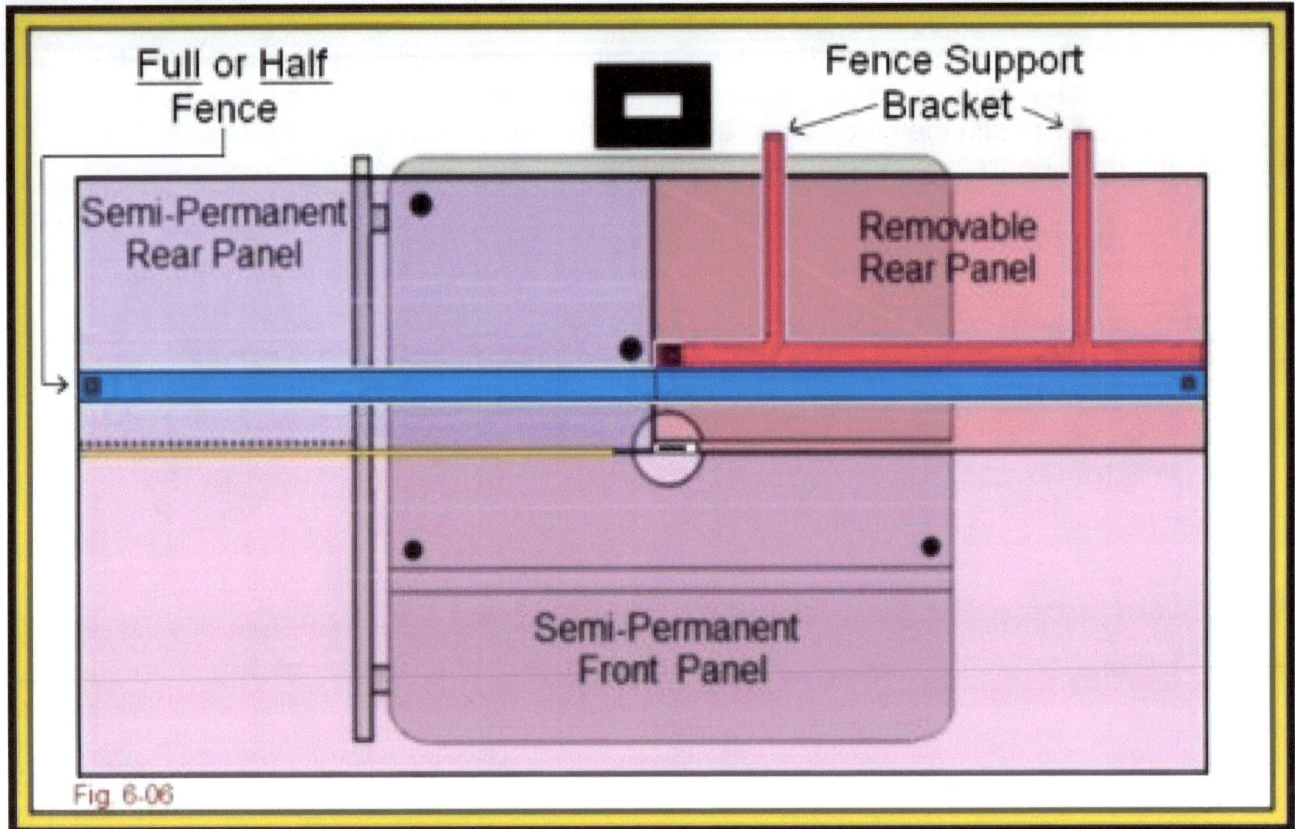

Fig. 6-06

At the core of this system is a three-part table configuration that provides for a long smooth feed-supporting Front Panel of whatever length you desire and a semi-permanent rear quarter-panel for leveling off-cuts with the front panel; both of which are simply bolted to the existing table with flat head machine screws. The *removable* rear quarter-panel is held down and in place by the Fence Bracket's attachment to the existing blade guide extension bracket and does little other than level the other two panels and provide for zero clearance when the fence is not in place; otherwise it can be clamped or semi-permanently screwed to the existing top. I used 1/4" plate for this and skirted it with 3/16" 2" x 2" for rigidity but one could as easily use 3/16", 3/8", or 1/2" plate. The separation of plates behind the blade (and more importantly: **in line with**, the blade) allows for insertion of a vertical stub fence; the purpose of which is to replicate the plane of the fence-before-the-cut, but from inside the kerf.

Details for building the fence itself can be found in Chapter Five but the adaptation of the resaw table conversion to traditional cast-iron tops is so subjective and the options so numerous that I'll only point out its salient features here. The yellow line in the above drawing is meant to represent a vertical strip of good, hard stainless steel about as thick as a bandsaw blade. It is screwed to a 1/2" x 3/4" aluminum bar under the table-top, which is mounted flush with its edge. Stainless is exceptionally rigid stuff and could be expected to remain vertical, without support, for half the length of a typical fence. The countersunk mounting holes of the stub fence's table plate should be slightly elongated (with a round file) to allow micro-adjustments of the stub fence to line up with minor changes in the blade's position. If one were to make the conversion-top with 3/8" or 1/2" aluminum plate, the stub fence could be screwed directly into the plate itself. I cut the stainless plate myself with a Dremel cut-off wheel but it was extremely tedious and you may want to see a machinist about that. You won't need more than two holes in each top plate to make it very secure. Be sure to de-burr all of these drilled holes, top and bottom. The flat heads on the top surface want to be as flush as possible and as far away from the plate joints as structure allows. I would drill the cast iron top from the bottom to insure bolt-ability: the ribs, you know.

Considering that the entire conversion-top is removable with four screws, it's your choice in how to deal with the conflicts arising from the zero-clearance lay-out shown here and the traditional key hole made necessary by the tilt options of a trunnion-mounted top. If one were to elaborate on this conversion to the point of building legs to support the extremities one would be almost to the point of re-building the whole saw; and that wouldn't be too difficult either. I do it all the time: and I'm a woodworker, like you. Just take some 1/4" aluminum plate and cut it out with your jig saw to look like the drawings below, have the two 4" I-beam pieces welded or screw them on yourself (edgewise), and you'd have a state-of-the-art frame on which to hang all the other little parts you need to make a blade go zzzinggg! The wheels, bearings, motors, and pulleys are all available on-line and your local tradesmen friends will fill you in on the rest. Go for it! You could build a $12,000 industrial-sized monster for $1,000 and some spare week-ends.

I drew the table-top in and colored it purple but you should think of the blade transport systems outlined above as component parts of a total band saw. The lighter you can construct the blade transport system the easier it will be to hinge and/or re-configure into optional material transport systems or table bases. If it's light enough it could be made portable, but by keeping it within reason the saw *head* can be used in a tilting head bevel saw. The most difficult parts to build are going to be the bent legs of the I-beam frame. There are several ways to do that and the best approach will depend on how sturdy you want to build it. In the direst of circumstances, one could wrap 1/8" plate in multiple layers to any thickness by applying the invisible screw technique I'll describe in the next chapter; it's labor-intensive but doable for the average garage-mechanic. If your local metal workers can't bend it to your plan with their brake, that's your ultimate fall-back method. You may find you'll have to break the outer, triangular wrapping into two or three segments to facilitate braking thicker plate.

With $200 worth of material and $300 worth of welding charges you'd be into such a frame for five or six hundred dollars but considering the advantages it would have over any other band saw made it would pay for itself in production. Good, well-made, die-cast aluminum wheels can be bought for about $60 apiece and you'd end up spending $300 more for bearings and miscellaneous hardware supplies. One could spend a year building a project like this and, in the process, make friends with every tradesman in your area and become the envy of Do-It-Yourselfers everywhere. The only dimensions I'll suggest are to allow 3/4" between the wheels (of whatever size) and the inside of the wheel housings; allow the same clearance to the frame (for co-planarity on a possibly warped frame); a full inch to the TT wheel's inner frame member; and 3/4" from the blade to the inner vertical frame member. I strongly advocate a third wheel, with the TT assembly mounted thereto, for its throat size and over-all efficiency and capabilities, but that, too, is discretionary. If you go with big wheels you should beef up the frame accordingly with thicker plate. Test each component as you go along and ask yourself if this frame will bend with five hundred pounds of spring force applied to it or if this lever arm will bend when forced, etcetera. Whatever size blades you want to run, the saw's components should be proportioned to that level of strength, without over-building; you still want to be able to throw your saw on a truck and go to the job site, right?

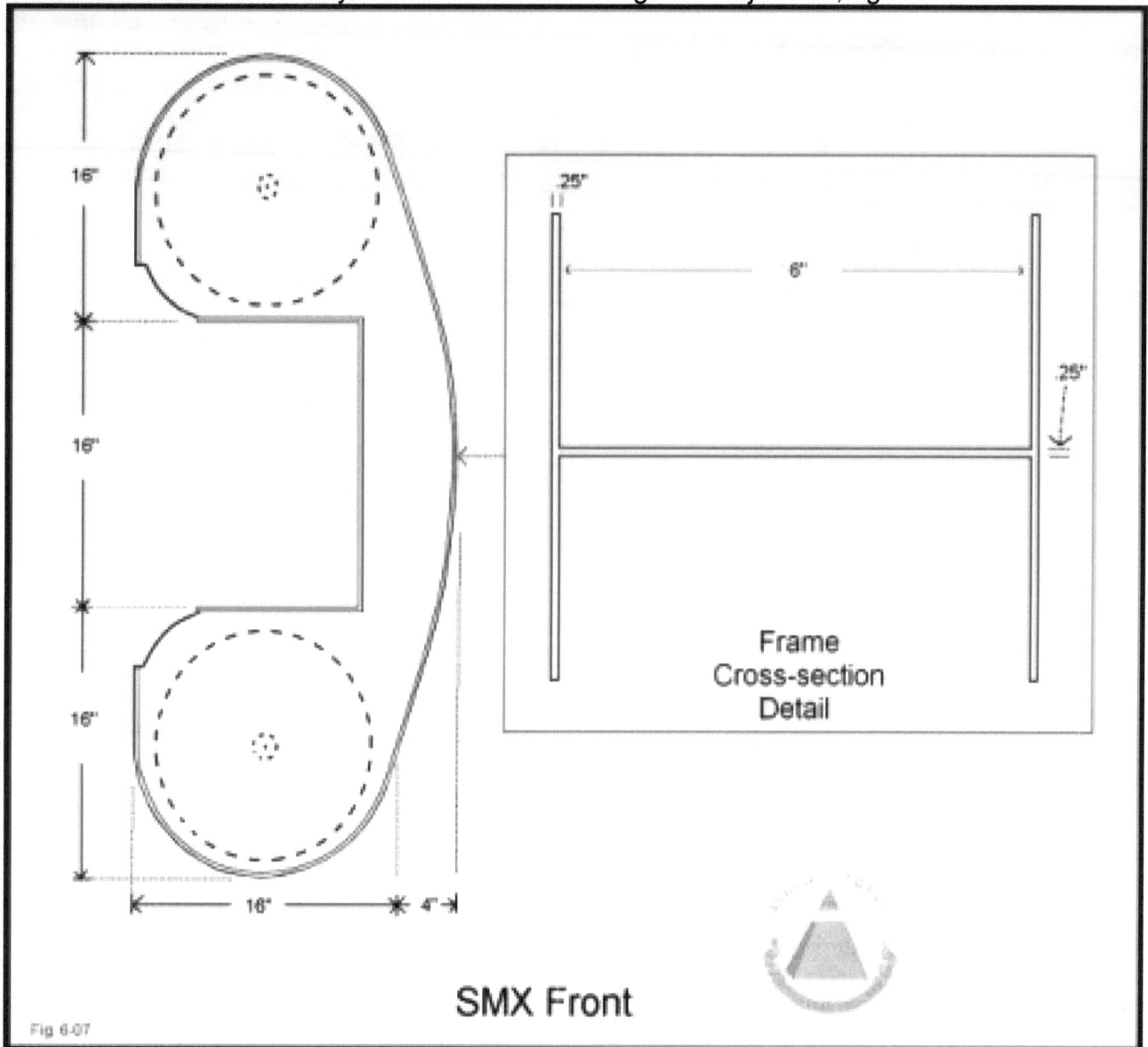

16"

16"

16"

16"

16" * 4"

.25"

6"

.25"

Frame
Cross-section
Detail

SMX Front

Fig 6-07

Now that you've got the entire saw built of aluminum and you can drill, tap, saw, and bend whatever you want into it or around it, you might as well deal with the three-handed veneer feed problem while you're at it. You could build a cheap and elementary hand-powered feeder that mounts adjustably right onto the table. You'd have to provide for twelve or more inches of slide-ability to accommodate widely varying workpieces and the addition or subtraction of wagon wheels with a snap-in/snap-out release bearing of some sort on a keyed shaft, which wouldn't be difficult with factory-made parts from MSC Industrial Supply or the like. Inflatable dolly tires are recommended but ultimately optional; and a spring-loaded tensioning system of some kind, also optional, would be a "plus" and very smooth, from the operational standpoint.

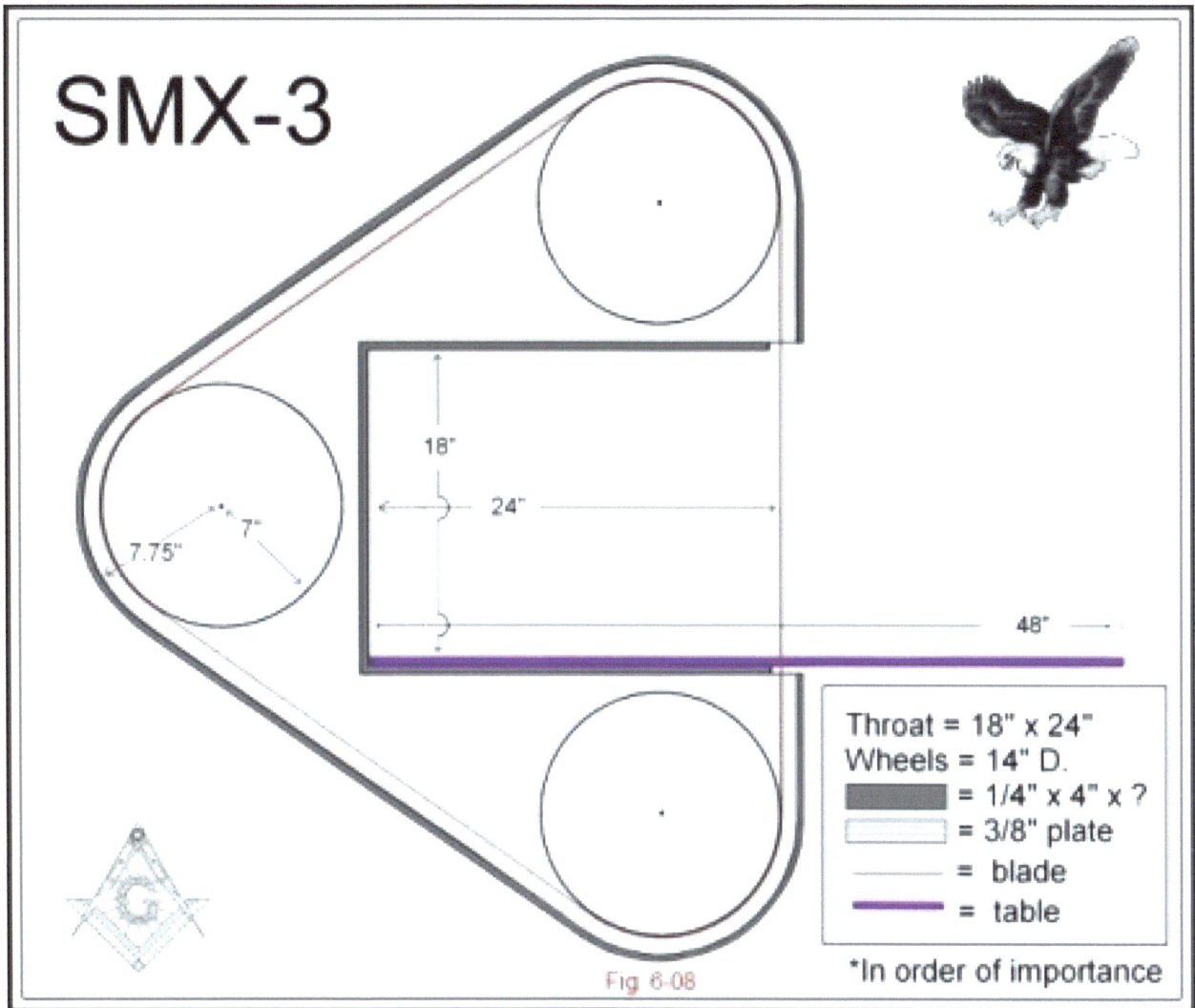

SMX-3

18"

24"

7"

7.75"

48"

Throat = 18" x 24"
Wheels = 14" D.
■ = 1/4" x 4" x ?
□ = 3/8" plate
— = blade
— = table

*In order of importance

Fig 6-08

The feeder below obviously wouldn't work as shown, but the concept is easy enough to understand and I leave it to the reader to round out the basic design details with a box to house the gears, a frame to fit the individual mating requirements of your particular saw, and a means to attach the top slide rail. I use the slide rail for the feeder extending inward to compose the top support-rail for the fence as well; serving both purposes at once. Projects like this call for a little research into parts and their descriptions. You'll find yourself in pages of the hardware catalogs you've never been before. They use strange new words and abbreviations there but they're usually quite well defined in the intro to each section. Some reading is required but once you see the patterns you'll be skimming through to find exactly what you were looking for and have it to your door on a truck the next morning. It never ceases to amaze me! It amazes me that more tradesmen aren't building their own machines; it's so easy now-a-days! The hardware is all there; just waiting for you to build the framework. Beware however; you could easily get hooked on such empowerment and find yourself in another business altogether!

I'm surprised such improvised crank feeders aren't already marketed as after-market accessories to bandsawyers for general veneering use. While the more expensive industrial auto-feeders save a bunch of labor, I prefer the hand-crank method for its "sensitivity" to feed pressure, its simplicity, and its ability to respond instantly to changes in the kerf, such as knots. There are numerous ways one could provide an optional source of power to the crank shaft however, and I see no reason it couldn't be made to convert as conditions change. With the use of v-belts and pulleys the crank shaft could even be rigged to accept rotation from a clutch-driven hand drill, with its built-in torque-sensitivity. Decisions; decisions.

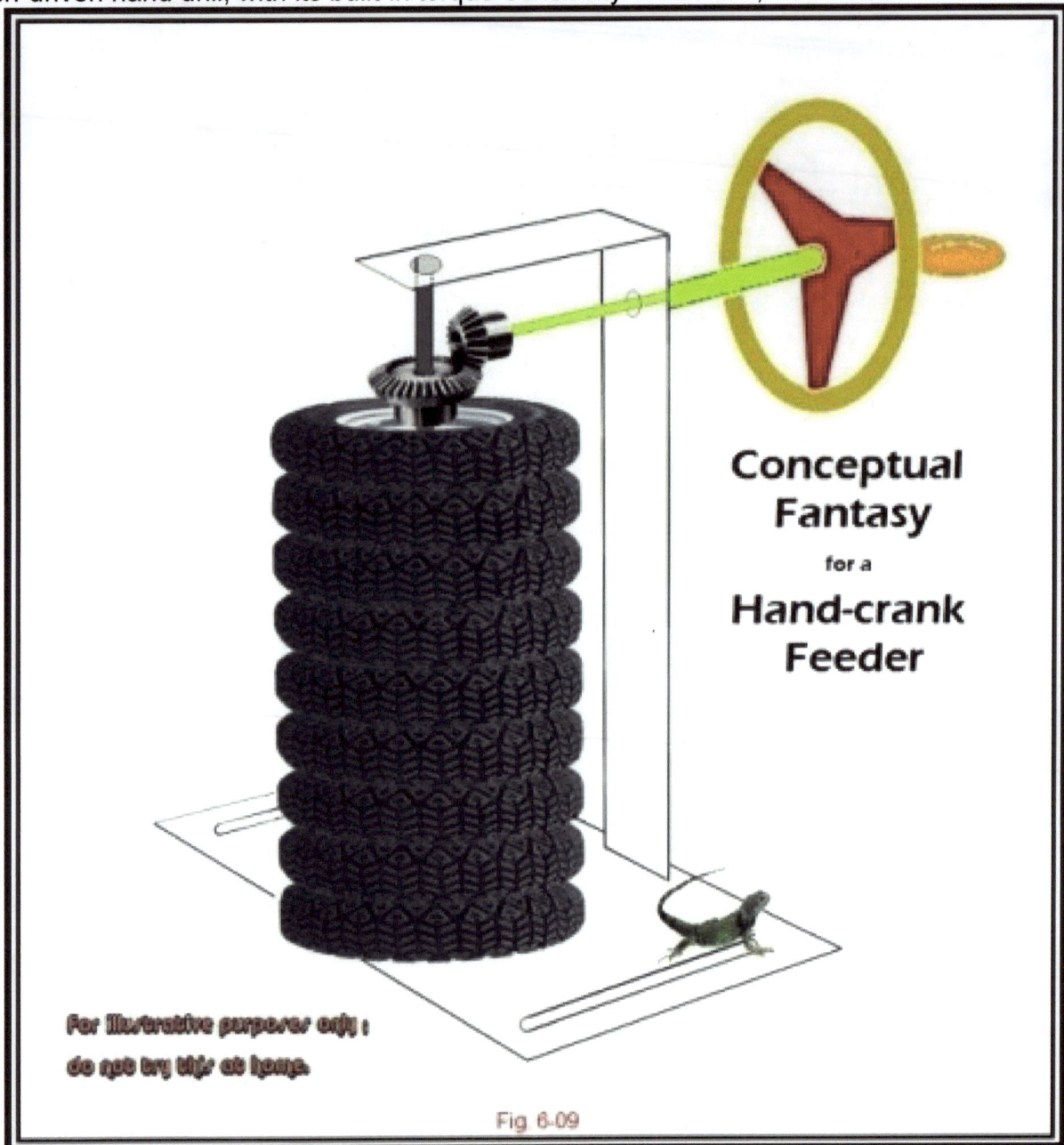

Conceptual Fantasy

for a

Hand-crank Feeder

For illustrative purposes only ; do not try this at home.

Fig 6-09

By now you're thinking I'm quite mad and asking: "If all this was such a great idea how come nobody's done it already?" Glad you asked. There are several good reasons. Back when bandsaws were a futuristic new concept they didn't have blades; they had to make them in a forge; and they weren't cheap; nor were they pretty. They were industrial tools, not available to the casual user. They required massive cast iron frames. The blades were dull. They were powered by steam. They were bullies.

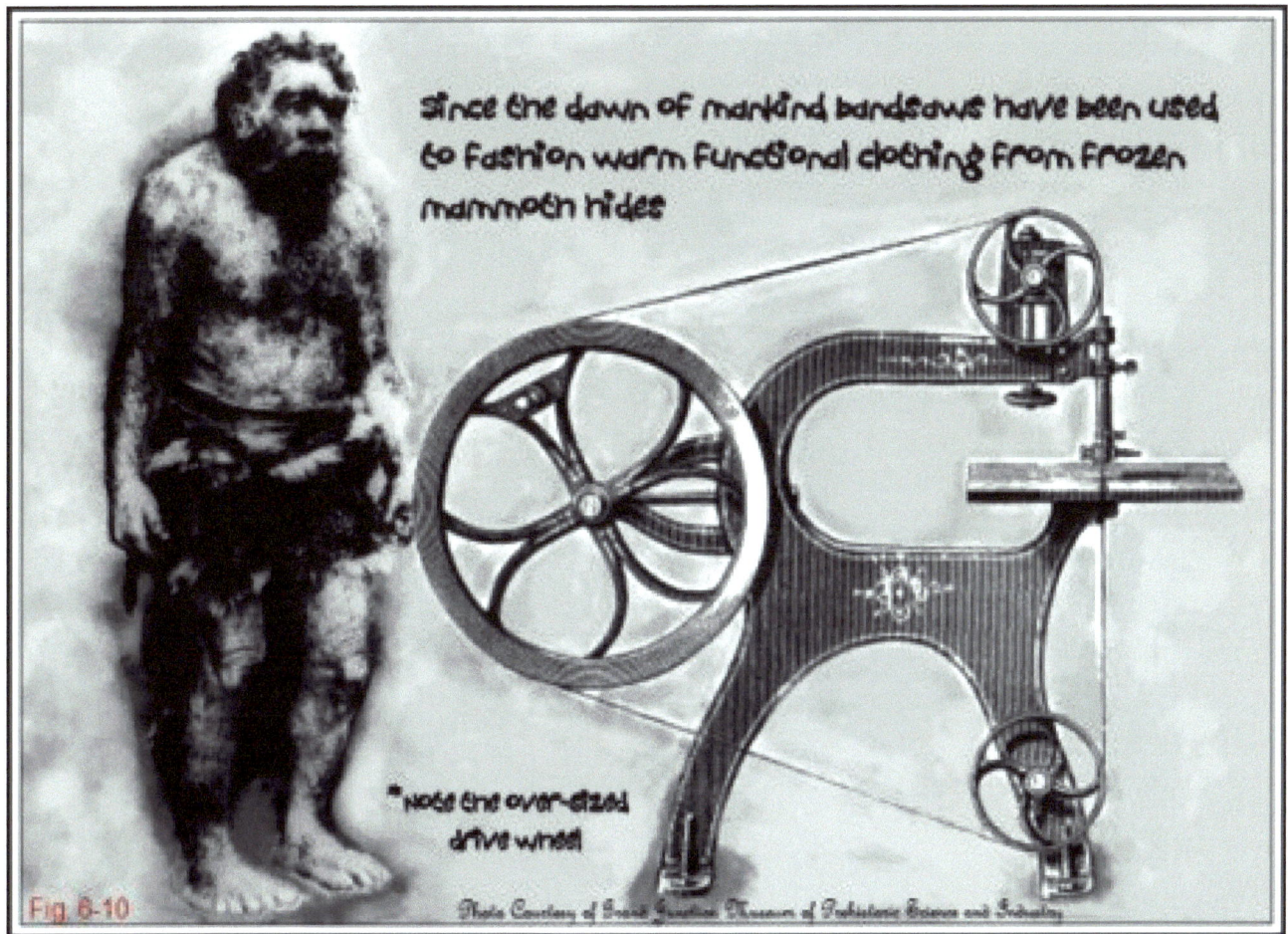

since the dawn of mankind bandsaws have been used to fashion warm functional clothing from frozen mammoth hides

*note the over-sized drive wheel

Fig. 6-10   Photo Courtesy of Grand Junction Museum of Prehistoric Science and Industry

Their operators had no shoes.

There are a ton of acme threaded rods, gears, and bearings to draw from and they're easily available at places like MSC and Grainger. The bevel saw below was designed and built in a month. It rips angles from 5 to 50 degrees and has a 24" (D) x 9" (W) throat. The 32" x 72" table top slides on eight nylon tracks and is guided laterally by radial ball bearings. It operates by turning the small spinner crank. The saw head is operated by the larger crank and locks in place with a knurled knob screw which tightens against a banana-shaped slide in the base frame. By fixing the fence to a movable table-top we avoided the complications of fence adjustment and a long slot for blade clearance. As the blade angle changes, the table slides back and forth to intersect the blade's junction.

Using this same idea one could effect any number of tilting head combinations to accommodate 0* - 45* bevel rips at any height relative to an adjustable , modular, and vestigial fence. The table itself, being segmented or modular, then becomes a platform for multiple feed and transport jigs. Once you get the weight of a bandsaw frame down to manageable proportions and are able to free yourself from the limitations of a table trunnion there's a whole new world of custom table sizes and configurations to handle those varied woodworking projects. That is why rational proportioning is critical to versatility. Treating the blade transport system as an independent component of the band saw frees one to build-in as much material transport versatility as one can imagine. With the tools and materials we have at our disposal today we can build bandsaws light enough and strong enough to serve as add-on attachments to much bigger tables than previously dreamed of. The table-tops can slide independent of the saw and the saw heads can tilt to bevel any angle you want.

Tilt-Head/Sliding-Table
Bevel Saw

Fig. 6-11

 Designing your own bandsaw would logically start with determining how much production you expect to achieve and how fast you want the saw to cut. Faster cuts rates will require wider blades which will require more tension, stiffer frames, and more power. To some extent, wider blades will also favor bigger wheels. There are no magic numbers but adding a third wheel should at least double the area (square inches) of throat and require a thirty-three percent stiffer frame. Starting with a known value of 150 pounds of spring tension to adequately run 3/8" blades, each incremental step upward should be accompanied by a 75 pound increase in spring pressure; roughly. As spring tension increases, so should the strength of the tension assembly and all its components; that means spreading the weight load over a wider area of the tensioning wheel's backplate and rolling the load on radial bearings. At some point, as you approach the status of a logging saw mill you should consider hydraulic tensioning. At that point you should also consider copying any of a number of well-designed saw mills that have already been developed. That's really outside the scope of this discussion.

Fig. 6.12

Times have changed since the first Delta 14" bandsaw rolled off the assembly line and we're no longer bound by the same limitations that craftsmen faced back in the twentieth century. The easy and widespread availability of parts and materials over the internet is already ushering in a flood of new products by and for woodworkers. We're already starting to see a surge in the number of after-market accessories pertinent to the field of bandsaws but, to date, they've concentrated primarily on enhancing the functionality of out-dated machines but virtually nothing has been done to up-grade the machine itself. Nothing has been done to expand the scope of our choices in regards to tilting tables versus larger, more functionally supported configurations.

Many behemoth-saw owners already have the components of a saw-mill (minus the depth of cut) and need only to cut off the base and plant the bare frame on horizontal tracks. And that's an option, too: I did it to a Delta Model 28-560, successfully.

Not many woodworkers are aware of this but intellectuals such as myself have known for centuries that bandsaws were originally three-wheeled machines and it wasn't until the nineteenth century, and the advent of the steam engine, that the third wheel faded into temporary obsolescence. It wasn't until the invention of the electric motor in 1953, when bandsaws first became available to the general public, that its design evolved into the familiar two-wheeled configuration that is commonly accepted today as the ultimate perfection of its design. Originating in the livery stable of Jebediah Lipschuitzer in Pooftersfroth, Wyoming during the long, snow-bound winter of 1949, the seminal prototype of an affordable band saw for the common man was first conceived. It was publicly introduced and acclaimed at the County Fair the following spring. Word of this breakthrough reached Washington, DC and fell on the ears of certain un-named Illuminati who, posing as Freemasons, marketed the concept under a logo suggesting Holy inspiration. That original marketing campaign, in fact, was so successful that its allegations are still accepted as *Truth* by the uninformed masses today. You can just hear old Jebediah rolling in his grave, laughing; his saw was only meant to cut outhouse-seats. Legends don't die easily however and Delta clones still dominate the windy plains.

Fig. 6-14

Does anyone remember why tables were made of machined cast-iron instead of the lighter and obviously less expensive steel plate? Does anybody know how tonnage came to be known as a virtue in regards to bandsaw design when the massive weight of these machines has been nothing short of their worst drawback? It's only been a few years since welded steel frames were introduced, but there are still those who would argue that the cast-iron frames were better.

There's an old joke: "How many bluegrass musicians does it take to change a light bulb? Four: one to change the bulb and three to sing about how much better the old one was."

Fig. 6-15

FIRST & PRYIBIL'S PATENT BAND SAW.

What was truly lost in modernization was the beauty and craftsmanship displayed in the castings they designed in the last century. It seems to be a lost art now and the "old arn" will always be admired for that. I can think of no reason a tool shouldn't also be a work of art.

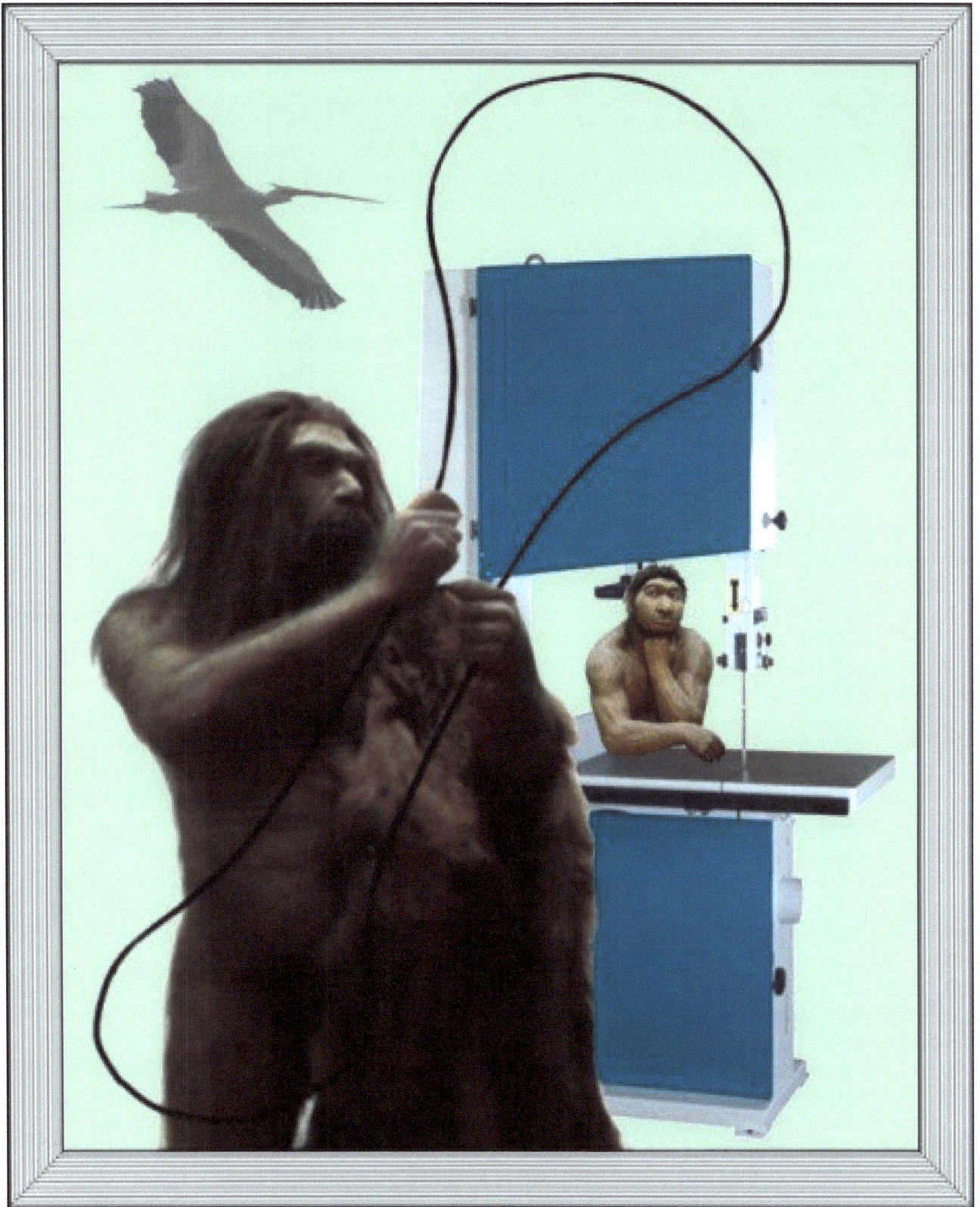

Fig 6 10

TOP Of The Line, or END Of The Line ?

It's plain to see that the saw above was designed to rip though ten or twelve inches of something under the greatest force possible with the most steel on the shortest table that would fit in a walk-in closet. A lightweight modern three-wheeler could do the same job faster with a fraction of the weight, footprint, and horsepower. If we presume there was an infeed/outfeed table we're left with the question of how the blade was changed. The avoidance of three-wheeled designs hits the brick wall of absurdity when ceiling height becomes a critical factor and you need a step-stool to reach the table. It's time for woodworkers to take the lead in re-engineering the tools of our trade; most of us only have a two-car garage with a low ceiling.

It is truly time to re-think the entire concept behind driving a long thin band of steel through a wider variety of ever-more-precious hardwoods. It needs to done from the ground up, starting with the choice of its composition. The materials available now were almost unheard of back when Jebediah was cutting thrones but you can be sure he'd have chosen aluminum if it had been available. We've come so far with the multitude of extrusions and weld alloys that it would be a snap for the average tinkerer to build a better saw. The tool steel parts and die cast wheels are already available on-line and the aluminum is in stock at whatever dimensions you need. All that's lacking is the will to do it.

It was my original intention after the Titan was "productized" to develop a whole new line of job-specific bandsaws and after-market attachments to serve the various specialized needs of the woodworking community but I see now that there's not enough time to go after all of them. Each frame, table, feeder, jig, fence, blade-setter, tensioning assembly, and feed-rail would take at least a year to test and de-bug. My intention now is to "get 'er done" anyway; even if I can't do it myself. What I can do is pass the know-how along and hope the rest of you will carry on.

If everybody was to start building their own bandsaws, to their own specifications, I bet we'd see a wide range of diversity in their design. It would be interesting to count the number of sawyers who thought the trunnion wasn't worth all the limitations it imposed and deleted it in favor of a removable beveling jig. I think we'd see more light weight resaws with long modular tables and high modular fences. I think we'd see more power-house three-wheelers with massive aluminum frames driving 1" carbide-tipped bands through 18"H x 24"W throats with a 2HP motor and thinking it was a table saw (sans kick-back). It would be interesting to see how many treat the frame as a separate, removable, and portable part of the saw table. It would be interesting to see how many saws weigh more than 200 pounds.

Re-introducing rationality and proportion to the design of bandsaws will no doubt lead to a refinement of the terms we use to describe them and a new vocabulary to classify them in terms of their specialized functions. I predict that at some point *spring rate* will become the dominant indicator of size, weight, and expense since it is tied so directly to blade width and frame construction. **Throat dimensions** shown as "24"(W) x 12"(H)" would indicate you're using your saw as a safer table-saw-substitute while dimensions like "12"(W) x 24"(H)" would indicate you're more interested in re-sawing. Other qualifiers; such as table dimensions, size and number of wheels, metallic composition, and horsepower would still be significant descriptors, but of lesser importance.

Remember, when you're looking at my portables, that their primary design goal was to keep the weight down to 75 pounds or less. While their throats are huge, their frames only support 150 pounds of tension. If you want to use blades wider than 1/2" you'll have no such weight restrictions and the SMX drawing shown in Chapter Seven can be modified to whatever strength your plan demands by thickening, widening, or bracing variations of the same basic I-beam, arched frame. If it ended up weighing 300 pounds without the motor you'd still be sporting a custom band saw of *Industrial* proportions.

Considering the incredible diversity of functions at which the band saw concept excels like no other, the *sameness* of consumer bandsaws available today is even more incredible. If we've seen one; we've seen them all. The Delta clones do an admirable job as middle-of-the-road hobby saws but totally fail to address the specialized needs of woodworking professionals whose design choices are becoming ever more restricted by the limitations of their bandsaws. Most of them accept this as a fact of life by saying: "If it ain't broke; don't fix it." But some of us do not accept this as a fact of life, and it is to them that the next chapter is dedicated.

In regards to all these conceptual drawings: if I had the time to develop every jewel of intellectual property my fertilized mind out-gases daily, I would. Given a short attention span, limited lifespan, design-to-market development lag, and a fervent desire to get out of the shop for air; there's more projects in this chapter alone than are likely to get done by me. Can you help out a little? It's your concern, too, right? With the easy availability of aluminum it's become possible, for the first time, for woodworkers to make first-class, state-of-the-art prototypes from the frame up, using the tools and equipment already in their shop.

CHAPTER SEVEN
Cheap Tricks

If the average woodworker knew how easy and fun it is to work aluminum, we'd have slicker jigs and a lot more mixed-media furniture. Working with wood is infinitely more difficult than working with aluminum. Aluminum doesn't warp, shrink, or split overnight as it dries; nor does it absorb moisture. Aluminum, with the slight exception of a few differences in alloy, is pretty much the same composition and reacts to tools consistently regardless of thickness. Aluminum can be surfaced with whatever laminate you desire for a finish, including wood veneer. I haven't done much with wood veneer laminations on aluminum but I can say with the voice of experience that the Wilsonart laminations I use on my saws and tables have withstood the industrial abuses of construction for ten years now without damage or loss. How wood veneer might fare would probably depend on the species used but the application of Weldwood contact cement seems to handle a relatively broad range of expansion coefficients. I see no reason why a skilled craftsman such as yourself wouldn't consider aluminum as a substrate in any veneer application. Given the ease with which structural elements can be joined with *invisible welds*, I'm surprised we don't see more wood-veneered aluminum cabinets and furniture being produced. As a design concept, this alone should inspire one to learn some basic metal fabrication techniques and lead one to at least experiment with mixed-media.

The skills required to work with aluminum are very similar to the skills you use working with wood and, with few exceptions, your skill in woodworking will show itself in the quality of your metal-work. They're both about shaping and fastening; design and finish. You might be surprised to learn that it doesn't cost near as much to tool-up for aluminum working as it does for wood-working at a professional level. If you're a professional woodworker now, you're only a couple hundred dollars of tooling away from being a professional-grade metal-worker. Limiting one's self to a single medium seems like such a waste of talent for the majority of professional woodworkers who might otherwise be master craftsmen in the broader sense. I see lots of mixed-media woodworker/craftsmen but there's a distinct disconnect when it comes to incorporating metallic substrates to wooden surfaces based, I believe, on ingrained mis-perceptions about working with aluminum. I can't know precisely what said mis-perceptions you might hold in your otherwise-enlightened cranium, but I'll venture a few that I most often hear from life-long veteran woodworkers.

If you're anything like me, as a woodworker or general contractor, you know every lumber yard and hardware store in a hundred-mile radius but I'm betting you don't know the location of a single welding or metal supplier. That's OK, I didn't either. They're everywhere; and if you don't have one nearby, there are many internet sources for aluminum as well. It's better to get the *heft* of the various materials and alloys by visiting the yards personally, however, and poking around in their racks if you can. If you enjoy scrap yards like I do you'll get all kinds of inspiration from just seeing all the possibilities in the aluminum pile. To my way of thinking there's no such thing as aluminum scrap and if I don't create enough scrap of my own I'll buy someone else's scrap to make up the difference. I often find special-ordered extrusions or extremely thick pieces in sizes more appropriate to my scale of metal-working that you wouldn't find in catalogs. To one who is even barely knowledgeable in the art of cutting, fastening, and finishing aluminum a visit to the scrap yard is like shopping in a toy store. It's like playing with the biggest erector set in the world. The possibilities are unlimited.

I was a general contractor when I started building band saws for a living. The only tools I had at that time were oriented towards construction so my learning curve was pretty steep when it came to working with aluminum. After ten years of manufacturing aluminum band saws I still have pretty much the same tools I used then. Besides having four Bosch jig saws and a Ryobi clutch driver now, the only new additions, tool-wise, are in the small boxes of bits for the routers, drills, and grinders I already had. What I knew about woodworking then still applies to the working of aluminum and with a few obvious exceptions, there wasn't **that** much more I had to learn to make the transformation from contractor to manufacturer. At least not so much that I can't teach you, in one chapter, enough to make adult-level aluminum gigs and fixtures for yourself. Hopefully, after reading this, you'll never again use plywood for your home-made fences, sleds, or shop jigs. There's no excuse for it and it tends to make professional woodworkers look one-dimensional, when I know that's not the case.

The first and most common complaint you hear from the *uninitiated* is that aluminum is "gummy" and clogs up on everything; but that's not true. It only seems that way because it doesn't transfer heat very fast and the heat generated by cutting melts aluminum at the point of contact. For some reason beyond my understanding of metallurgy, aluminum will braze itself to steel at relatively low temperatures and since aluminum doesn't dissipate heat as fast as we'd prefer; in that split second when the steel edge cuts the aluminum shaving there's enough heat built up to braze the two together; "gumming-up" on the cutting edge. There are several cures for this but wetting the tool with cool water is the cleanest of them all.

When I started out I was on the verge of terminal despair from all the broken jig saw blades and drill bits but then I experimented with using a constant stream of water as a lubricant. It worked like a charm and my aggravation disappeared after that. I thought I'd discovered the metal-working equivalent to a cure for cancer! When I told my welder about this incredible new discovery he replied by demonstrating how fast his jig saw cut using WD-40 and it was indeed a little faster but it left an expensive little puddle of goop on the floor. I've since learned there are other techniques involving grease and oil and even wax but they they all leave residue and cost money. I haven't tried most of them but I doubt they're meant for the heavy-duty, thick, prolonged cuts I do hour after hour in my shop.

My machinist, too, thought water cooling was the way to go and went so far as to rig a water-misting device to his air compressor and connected it to his jig saw for production cutting. My only reservations about the set-up he had was that the supply tubing kept his jig saw on a short "leash" while I needed to move around the shop freely with my jig saw to use the various vises and jigs scattered around the four walls thereof. Also, I don't have a compressor, because there's no place to put it. In the average woodworking shop, all things considered, I think it's far simpler to fill a Windex bottle with (free) water and spray with the left hand while drilling or sawing with the right. Drilling operations are usually best lubricated by frequent dipping of the tool's bit. I keep a bowl of tap water close at hand. Sometimes I dip and spray, both. One of my favorite things about working aluminum is that the shavings drop on the floor immediately and don't float in the air like saw dust. I can shop-vac immediately or wait until the water dries and clean up after lunch. I create a lot of water puddles with this constant spraying but the shop-vac cleans my parts as well as the floor with no residue. If you think woodworking is an environmentally clean occupation you'll really like working with aluminum. But while cutting and fitting aluminum is relatively clean work, grinding and polishing operations do send a cloud of dust skyward and I wear a mask when doing so. Since the dusty parts constitute only a small portion of the process I do it outside; and my shop is remarkably clean. Shop dust can totally ruin the appearance of a white lab coat.

Metalworking is typically associated with welders and machine tools of a nature totally unfamiliar to the woodworker and their presence in the average woodworking shop would be unimaginable to most of us but you shouldn't let that stop you from undertaking such projects as jig-making, furniture design; or in my case, building bandsaws. I saw a T-shirt at an air show in California that said: "There is no problem that can't be solved with the proper application of explosives". It was worn by a military demolition specialist of some sort and parallels the thinking of most tradesmen in that they view every solution to every design problem through the lens of their particular specialty; not the wider view of over-all quality. I've noticed that metalworkers, in general, prefer steel to aluminum for their pet projects and work it with the same techniques and tools they use for steel. Aluminum is expensive compared to steel, and they have a sizable investment in metalworking tools designed for steel. Because of its expense, aluminum seems to fill an unclaimed media-niche between woodworkers and metalworkers that few are eager to jump into, but all would do well to master: especially woodworkers, who rely so heavily on jigs and home-made shop tools. While the warmth and beauty of wood looks perfectly natural in the home; its relative instability, and the difficulties of joining it, make its use problematic in industrial environments. Deep down inside, you know those plywood shop jigs look like a third-grade science project when applied to a shop tool because they always look so out-of-place. No matter how skilled you are as a woodworker or how well executed your wooden jigs might be, you'll be judged as negatively for your choice of materials in these projects as you would be for building _____ out of _____ (fill in the blanks for yourself). You probably already know, instinctively, the circumstances wherein wood or metal is the appropriate media.

While welding may be the fastest way to join aluminum, it is not always the best way. Welding invariably warps the metal through heat expansion and subsequent shrinking, making the welded piece unsuitable for lamination. Welded aluminum isn't an attractive material in its raw form and even when it's polished it still needs a coating of some kind to prevent oxidation. Welded, cast, and machined aluminum are common everywhere and you'll probably find a custom metal fabrication shop near you for when that special need arises. For

the average woodworker's purposes, the few metalworking tools in the **My Favorite**

Tools

illustration below are enough to build everything you need for your shop and some elegant futuristic furniture designs featuring aluminum as a substrate.

Generally speaking, when aluminum parts have to be perfectly flat, square, or fit precisely, you can't weld your joints because aluminum warps so badly when heated. I discovered by accident, however, that once you drive a sheet metal screw into two pieces of aluminum they will never, ever, come apart unless you aggressively un-screw them. Once you grind the heads and tails off of said screw and un-screwing is no longer an option, you have to resort to grinders because that's the only way you will ever separate those pieces again. You'd think, soft as aluminum is, that a small sheet metal screw would pull out rather easily. You'd be be very wrong. You literally have to grind away all the aluminum surrounding a misplaced screw before it will let go. You can't tap it out, drill it out, or pry the pieces apart without first destroying both pieces. Whenever I have to separate two pieces thusly joined I grind the screw out with a small Dremel cutting disc, keeping the kerf as narrow as possible, then replace it with a 1/4" or larger machine thread. You can grind the head and tail off again and the joint is still invisible so all is not lost.

The slight indentation left by the grinder is usually visible after such operations but, if done carefully, becomes invisible again under a coat of powder-coating. If the divot is deep, lamination is sure to cover the tracks; but thin enamel or lacquer paints aren't recommended. If you plan to spray paint your project you'll have to exercise great caution when grinding the heads off of screws. I've found that the invisible *flush screwings* are achieved by the same tool that removes them fastest; the 4.5" flap-disc. You just do it more carefully. Watch the sparks; when they disappear you're grinding aluminum so lighten up and just touch the surface lightly with the flapper. When the screw-head is gone the frictional-drag of disc-on-aluminum will tell you immediately to back off.

Just A Few of

My Favorite Tools

Fig 7-01

The most often used joint in my shop is one I call the "invisible weld". I call these *pseudo-spot-welds* "invisible welds" because when you powder coat the assembled parts you can't even find their location even when you know exactly where they are. I use 1/4 - 20 tapped holes the same way for thicker plate but #8 sheet metal screws are perfect for joining aluminum sheet to anything of equal or greater thickness. If you'll turn to page 2587 in your Grainger hymnal now you can select the length appropriate to whatever your project requires. I keep a weighty stock of all sizes but the one that gets most usage is the 1NB64, (5/8"L), and checking my records I see that I've used over 10,00 of them in the past few years. This is the exact product description: (Self Drilling Screw, Sheet Metal, Hex Washer Head, Case-

Hardened Steel, Zinc Finish, Size #8, Length 5/8 In, Drive Type Hex, Driver Size 1/4 In, Width of Hex Head 0.250 In, Head Height 0.096 In, Washer 0.272", Drill Point Size #2, Drill Point Length 0.211", For Attaching Variety Of Materials To Metal Up To 12 Gauge, Meets/Exceeds SAE J78, 1310/Package) You'll need to use a 9/64" drill for these. A 1/8" drill will break the heads off before you can get it through. I don't know what a #2 drill even is. Forget it.

When you grind the heads and tails off you have to use an aluminum-friendly flap disc sander because the aluminum will clog the grit on normal paper-backed sanding discs by the time you get to the third screw and it becomes useless even when used against steel. I learned the hard way that ordinary grinding wheels and sanding discs won't work on aluminum, so save yourself the frustration; they "cake up" immediately, becoming useless (and you can't "un-cake" them either). There are several brands and styles of flap-discs but the ones that last the longest against steel screw-heads, remove aluminum the fastest, and leave the smoothest surface are the Westward Type 27 Flap Disc, (Disc Diameter 4 1/2 Inches, Grit Size 40, Arbor Hole 5/8-11 Inch, Maximum Speed 13,300 RPM, Zirconia, High Density), from Grainger: # 6NX83. CAUTION: I use this tool hour after hour on some days and have suffered some of my most grievous skin-gougings from it. You can be working along with both hands happily moving around your project when suddenly, without warning or provocation, your right hand will viciously attack your left hand in a brutal, senseless act of random violence.

For safety, you should always clamp your workpiece down when using angle grinders. I prefer to clamp my work to a movable table rather than a secured workbench because if things go horribly wrong, which they often do, it won't be **you** that gets thrown across the room, if you catch my drift. Angle grinder discs have three exposed working surfaces: one that pushes your tool to the left, one that pushes the tool into your vulnerable parts, and one that pulls the tool away from you. Never, ever, let a flap-sander disc snag on the portion of the disc that's coming your way. Do not remove the safety guard on these little demons, they can really rip skin. While not lethal, they can really hurt. Lots of blood. And the affected parts are nowhere to be seen - just red spray. Very gory. If you grab the tool firmly however, with both hands at all times, and follow this simple procedure, you have nothing to fear.

The same applies when using the abrasive discs - especially the black, non-woven Norton Rapid Strip 4.5" metal finishing, 5/8-11 threaded, silicone carbide, extra coarse discs (Grainger # 6PZ17). Said black Norton abrasive discs are even more aggressive at removing aluminum than the flap discs but they wear out very quickly if you subject them to steel screw heads or any other steel parts. For general surface smoothing, cleaning, and de-burring of aluminum they're awesome. They're good for rounding edges and smoothing out the dimples left from screw-head removal. Aluminum gouges easily and deeply in the supply yard and gets further abuse at your hands in the shop so it may need a once-over with the coarse polisher before laminating or powder coating. Gouges, if left on the surfaces, will show through powder coat and tend to snag on your dry-wall paddles as you're applying contact cement for lamination. A quick going-over with the coarse polisher will also remove any non-adhesable (I lovemaking up these new words) substances that might have gotten on your project, such as drool. If you plan on polishing the bare metal it's a good first step.

The next step toward the polishing of aluminum would be my other favorite abrasive disc: the Maroon Wheel (also a Type 27, depressed core). There are several styles and brands for these also but after likewise trying all of them, my favorite is the Maroon Wheels available nowhere else that I can find but here in Grand Junction at Bonner Supply, our local metal supply yard. (970-241-2551 - tell them Bill sent you) Be sure to specify the 5/8 - 11 unless you want/have the adapter. The threaded discs are much faster to change and when you're going back-and-forth between grits during the finishing phase it's aggravating to have to change mandrels every time. I have three DeWalt angle grinders (they get hot in your hands when you use them all day so it's nice to rotate them for a cool one); but you only need one. For the ultimate fine finishing of aluminum, such as it is, I use variations on the ubiquitous Scotch-Brite pad. They come in a variety of grits from any metal supply house.

Whenever you're drilling or tapping aluminum you should have a bowl of water handy to dip your tool into; do it often. If you're joining two pieces invisibly you'll need to clamp as close to the joint as possible because if the shavings can get into the space between them, they will; and you'll have an ugly gap that won't go away. Drill and screw one hole at a time starting from the outside of your array working toward the center and clamping each operation. Do it while the pieces remain tightly clamped so you're sure to have all the holes aligned perfectly. There's no tolerance for mis-aligned holes; the screw won't thread and you'll bind until the head breaks off; a major drag, that. You absolutely, positively can not have too many clamps. It sounds time-consuming but you'll be amazed how fast it goes once you're set up and get the hang of it. In fact, once you get the hang of it (and it ain't hard at all) you'll wonder why you ever bothered with wood. If nothing else, the ability to work with aluminum will open new doors for jig-making and your woodwork will show a marked improvement. Men will admire you and seek your aid in the art of jig-making; women will flock to your doorstep.

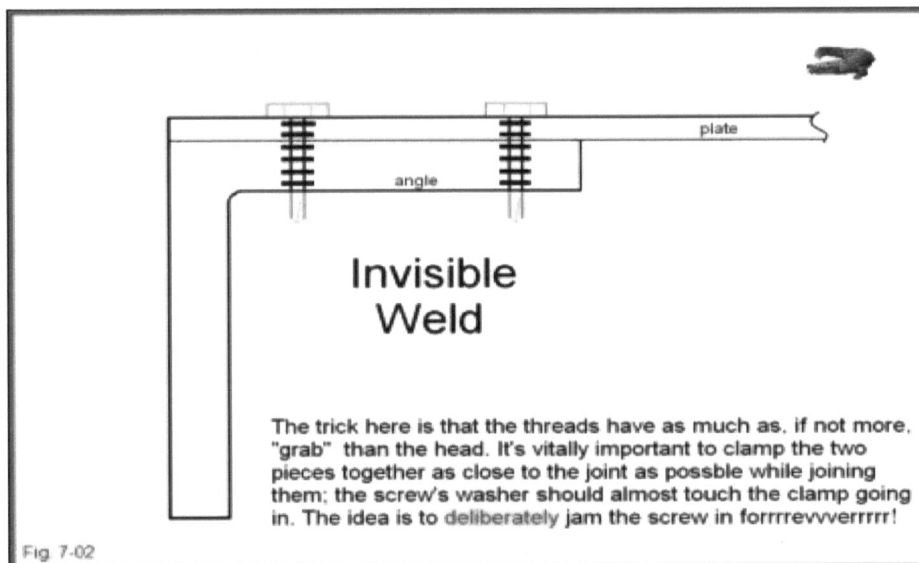

Invisible
Weld

The trick here is that the threads have as much as, if not more, "grab" than the head. It's vitally important to clamp the two pieces together as close to the joint as possble while joining them; the screw's washer should almost touch the clamp going in. The idea is to deliberately jam the screw in forrrrevvverrrrr!

Fig. 7-02

For joining thicker stock with 1/4-20 thread I drill 13/64" holes to their full depth; drill 1/4" holes through the top piece **only** (use assorted lengths of scrap to control it) and tap 1/4 - 20 threads all the way in: that way you can suck the pieces in tight together. For 1/4-20 taps going deeper than 1" , I drill 7/32" holes, because at that depth you still have enough "grab" and it's easier to tap; and tapping at that depth is really tedious. When tapping for 5/16-18 threads I use a 1/4" drill; for 3/8-16 threads I use a 5/16" drill. The counter-sink is also important; use only three-flute countersinks in AL; keep it wet (dipping often); and lean on it hard, don't let it chatter -dig in until you see **spirals** of aluminum shavings. As with drills, taps, and saws: check for aluminum build-up on the cutting teeth frequently. Scrape or pick it off if you have to and vow to use more water next time; then sin no more.

Working around clamps this close to the drill is often problematic and I use vise-grips whenever possible but it is sometimes preferable to use the 6", 12", or 18" aircraft drills.

Drilling and screwing at a slight angle is acceptable too. I never could find a 9/64" aircraft drill, so in a pinch, when there's no other way, you can use a 12" L. - 1/8" drill to wallow-out the hole enough to seat a #8 screw.

A Ryobi Clutch Driver will keep you from ripping the heads off if you set the clutch at 17 (*units of ?*). I pre-drill for #8 sheet metal screws with a 9/64" bit because pre-drilling 1/8" will bind the screw in more than 1/4" of depth, breaking the heads off the screws before they can seat themselves. It usually doesn't matter because I was going to grind the heads off anyway but I like to keep my options open. **Do not ever** grind the heads or tails off until you're absolutely, positively sure you won't have to take it apart ever again. I have much experience in these matters; none of them fun.

When I first started down this path to aluminum madness the only taps I had were hardware-store-stock, four-fluted, general purpose, taps. If that's all you have in your shop I genuinely pity you, for your aluminum-working experience is going to suck; horribly. Two and three-fluted taps will have to be ordered from a machine-tool supplier like MSC. They're not hard to find, easy to order from, and very helpful on the phone. I didn't know that then but my machinist set me straight. It makes all the difference in the world and I wouldn't even attempt tapping aluminum with a four-flute tap now. Joining two pieces with an "invisible" machine screw thread is pretty much the same as joining them with a #8 screw and you still need to keep the pieces clamped until the screw is in place; the least movement will result in a cross-threaded mess. Probably the only difficult thing about metal work is the selection and placement of clamps; and accumulating a large selection of them. I can't get past the $2 tool bucket in a hardware store without picking up at least four more clamps, often not knowing what I'll use them for. But lo and behold some day they'll be "just exactly what I needed". Likewise; you can't have too many vise-grips. Or too many 3/8" drills, extension cords, and plug-in outlets. Or money, for that matter. There's no end to it, is there?

Hot-rodded tools:

Exhibit # 1

I don't want me, nor my hands, near this beast when it's cutting so I clamp every little piece religiously. The duct tape covers a 1" hole where a small chunk of AL shot through the pot-metal safety cover and took out a light bulb across the shop. I wasn't hurt because I treat every tool like it's "out to get me". The point is:

The
Cheap
Chop
Saw
Conversion

"Jig Up For Safety"

Home-made clamping fixture with multiple taps for holding assorted workpieces. The table itself is tapped in many places to aid in clamping

1/2"
X
1.5"
Bar

AL bar is tapped to bolt through existing fence.

for extreme angles and odd shapes. The saw itself gets bolted the workbench as part of infeed/outfeed set-up.

Fig 7-03

This is the second aluminum blade I've had in my old Black and Decker chop saw. The first one lasted five years and this one is now four years old and showing no sign of giving up. It's a 10" - 80 tooth Delta blade made specifically for aluminum and cost about $80 back then. It can throw small chunks of off-cut aluminum three miles and has a muzzle velocity of 87,000 feet per second. Like any other strategic ordinance it's best to secure the area and stand back when firing it. It's never hurt me because I've never trusted it. I never get in-line with the plane of the blade when it's cutting. I clamp everything down religiously. I keep my fingers no closer than the trigger allows.

That said: it's really sweet for cross-cutting and squaring everything I do. It generates heat like no other tool in the shop and you can trash these blades in a heartbeat so you have to pump water into it as fast as your little fingers can squeeze that Windex bottle. You want to direct the spray at the center of the blade, just below the safety guard, at an angle, so the water shoots outward centrifugally from the middle. Doing that causes the water to spray into the kerf at high speed and wash around both sides of the teeth. It can cut a 3/4" x 4" thick aluminum plate in seconds but you have to take it slow when cutting thinner material, like

1/8"-wall tubing, because you don't want it to "bite" - ever! At 10,000 RPM a little common sense is in order here. Think your set-up over real good before you pull the trigger. In other words; it's no more nor less dangerous than your table saw. Always remember the water; lots of water.

Since there's no way to hold a workpiece with one hand, squirt with the other hand, and hold the trigger with your third hand; you have to clamp everything down very securely. Long pieces are easy to clamp but the little, short pieces present all kinds of clamping problems so I've got a whole box full of little clamping jigs that screw down directly to the table with all kinds of knurled-knobbed threaded rod screws of different lengths. You make them as you go along so save them because you'll probably need those setups again some day. It might sound like a lot of extra work to go through but mark my words - you'll use these techniques for all your other tools once you get the hang of making your own clamping jigs for this one.

Bear in mind that all of the following jigs were built on the spur of the moment with the scrap at hand and no thought was given to their beauty. In use, they're subjected to water constantly, as are the saws they produce. Working aluminum without welding leaves a remarkably clean shop. And it's really cheap to get plate sheared-to-size at your local supply house; plus you don't have to buy the whole sheet.

Speaking of clamping jigs, if you're doing any production work it would be hard to imagine making any money at it without a host of jigs. These are but two of my many assembly jigs and I have no idea how many thousands of hours they've saved over the years. It would be equally hard to imagine these jigs being made of wood. Wood just can't handle the constant beating these jigs are subjected to:

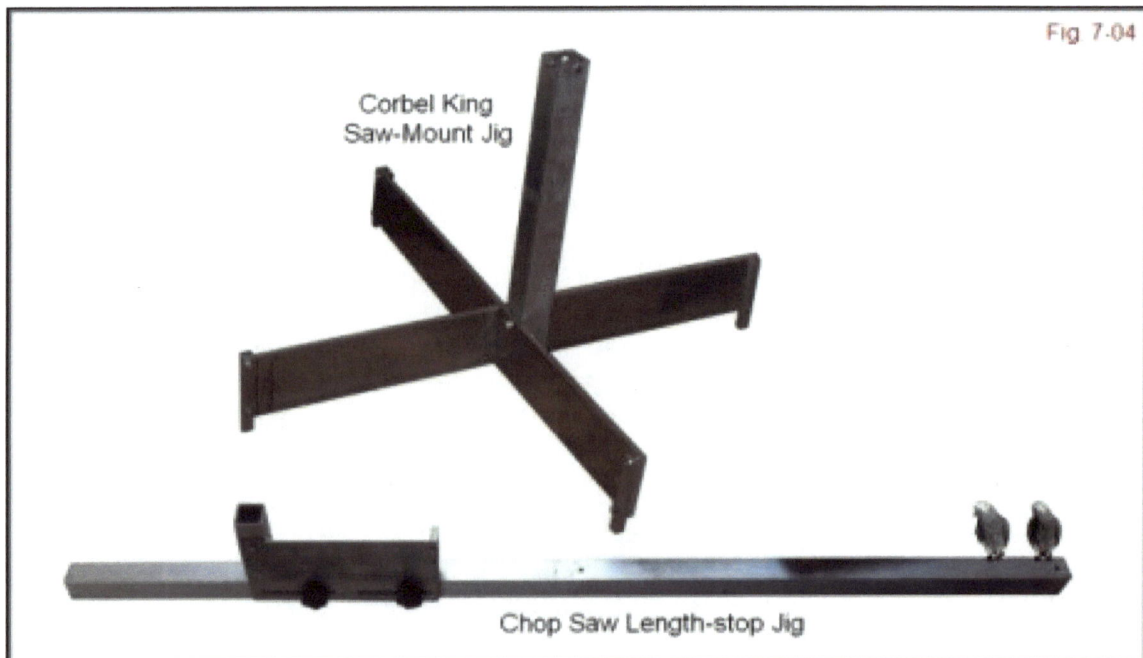

Fig 7-04

Corbel King
Saw-Mount Jig

Chop Saw Length-stop Jig

Granted these are metal jigs designed for metal assembly, but can you imagine applying the same precision to the wood products you produce in multiples. And the hours you'd save. It would be pointless to describe here the function each of these jigs serve, they're proprietary, but the ease with which they're made should be obvious. They're not warm, alive, or beautiful, but they are forever and the dimensions never change; they never warp, they never wear out. You can whip up an aluminum jig in an hour or two and never have to lay-out, square, measure, or clamp the same job again. If I'd known about jig-making back when I was working wood I'd have made some money at it.

Unseen in either picture are the 2" x 4" "legs I made out of square tubing that allow me to get my hands under it to attach nuts and bolts. Everything that comes out of these jigs is exactly the same as the other and they all fit subsequent jigs respectively. The parts that go into these jigs are exactly the same because they, too, were made with jigs like the two below. The *8-shaped* holes in the Titan jig allow access for the many C-clamps I use to pull the angle into its nominal shape. Holes in the Corbel King jig allow through-access for nuts and wrenches to tighten them.

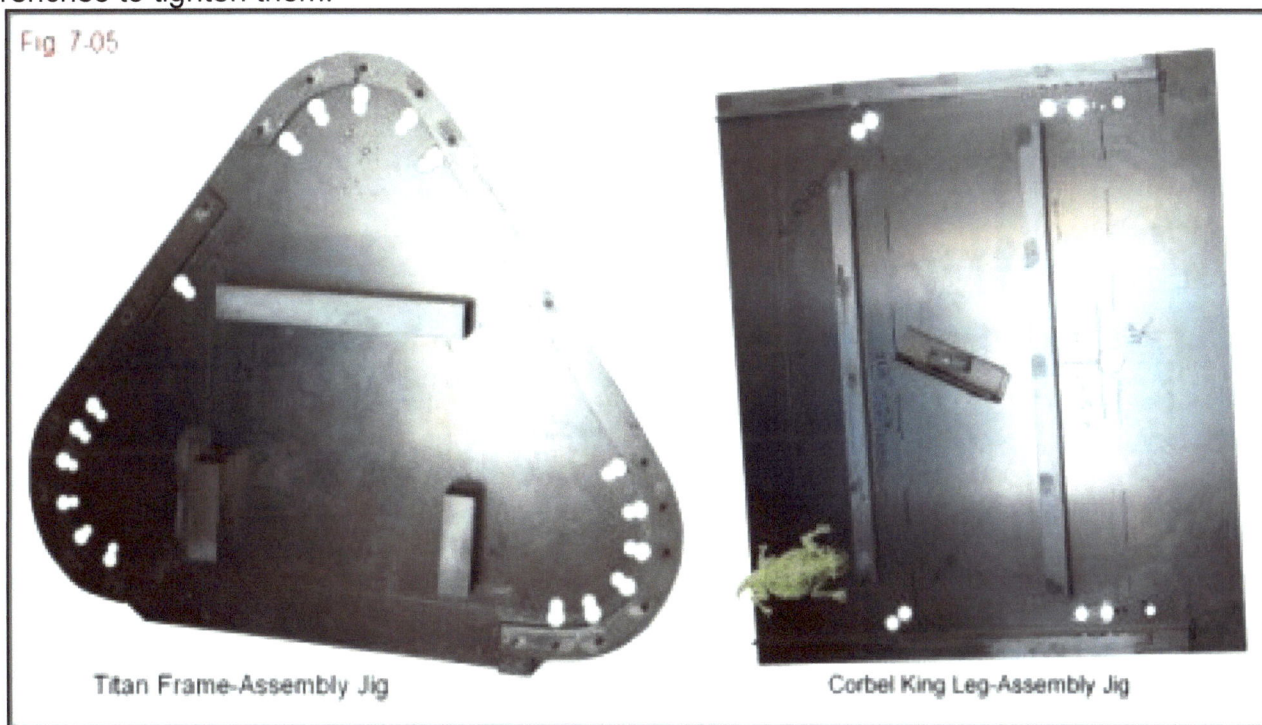

Fig 7-05

Titan Frame-Assembly Jig   Corbel King Leg-Assembly Jig

The bolt-head coming out the top of the Corbel King mounting jig above micro-adjusts the height of the Titan in relation to the table's top and demonstrates another of my favorite cheap tricks: the square-tubing end-plug. I have to weld the mitered corners of my saw frames because the options are too time-consuming but on all other tubing joints I use end plugs and invisible welds to connect ends. I keep a constant supply of tube plugs handy by cutting up all my 1/2" plate scrap in the magic chop saw.

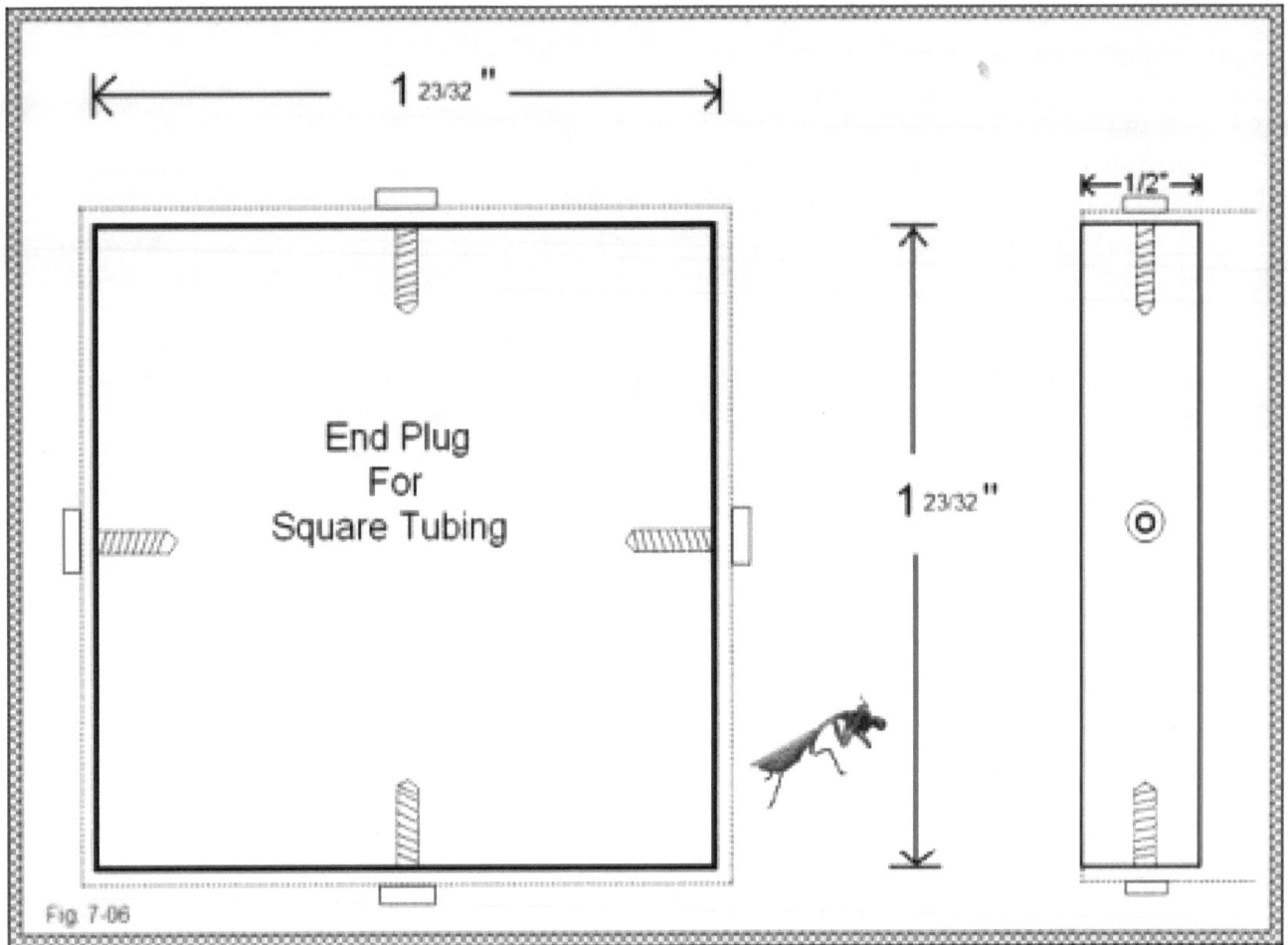

Fig. 7-06

Once it's set up to cut 1.71875" it's time to convert scrap into solid gold and I make as many as I have the scrap to process. End plugs give you the options to drill and tap into them from any, or all, of five different directions for connection to other pieces. The entire weight of one of my Titans is supported on the strength of two 1/4-20 flat heads tapped into one such end; and after ten years there's not been a failure. Aluminum in this configuration has so much flex you can't tear it apart and I suspect, in this application, that the brittleness of a welded joint would not fare so well. What I like best about this technique is that the junctions are square, straight, flat, and clean enough to butt any other square-cut component right up to it; including decorative laminates (making it something you should seriously consider, design-wise, for furniture applications).

These can also be done with 3/8" plate and lately I've been having my water-jet machinist add these little squares to his program and now all the scrap from other parts is automatically rendered into perfect little squares. I might also add that if you aren't familiar with your local water-jettist, you should introduce yourself. They can turn hundreds of dollars worth of aluminum into thousands of dollars worth of parts without really trying. Don't worry if your home made end plugs aren't perfect, they just want to be as close as you can get.

In this case it's a proprietary jig but the same idea would apply equally to the construction of a round or oval table-top. You would grind the screws down flush, apply the laminate of your choice, and trim it the same way as you would a solid or ply-wood top. The only real difference would be that the table is lighter and structurally stronger, without the bulk of a wooden table. Using thick plate you could rout an ogee edge while you trim. I use a round-over trim bit to dress up my Corbel King tables and I think the contrast between black formica, shiny aluminum, and hammer-gray powder-coat is cool. It's basically a whole new style that hasn't been explored yet but I intend to, when I get the time. Carbide router bits seem to work just fine on aluminum, but you have to cut a little slower.

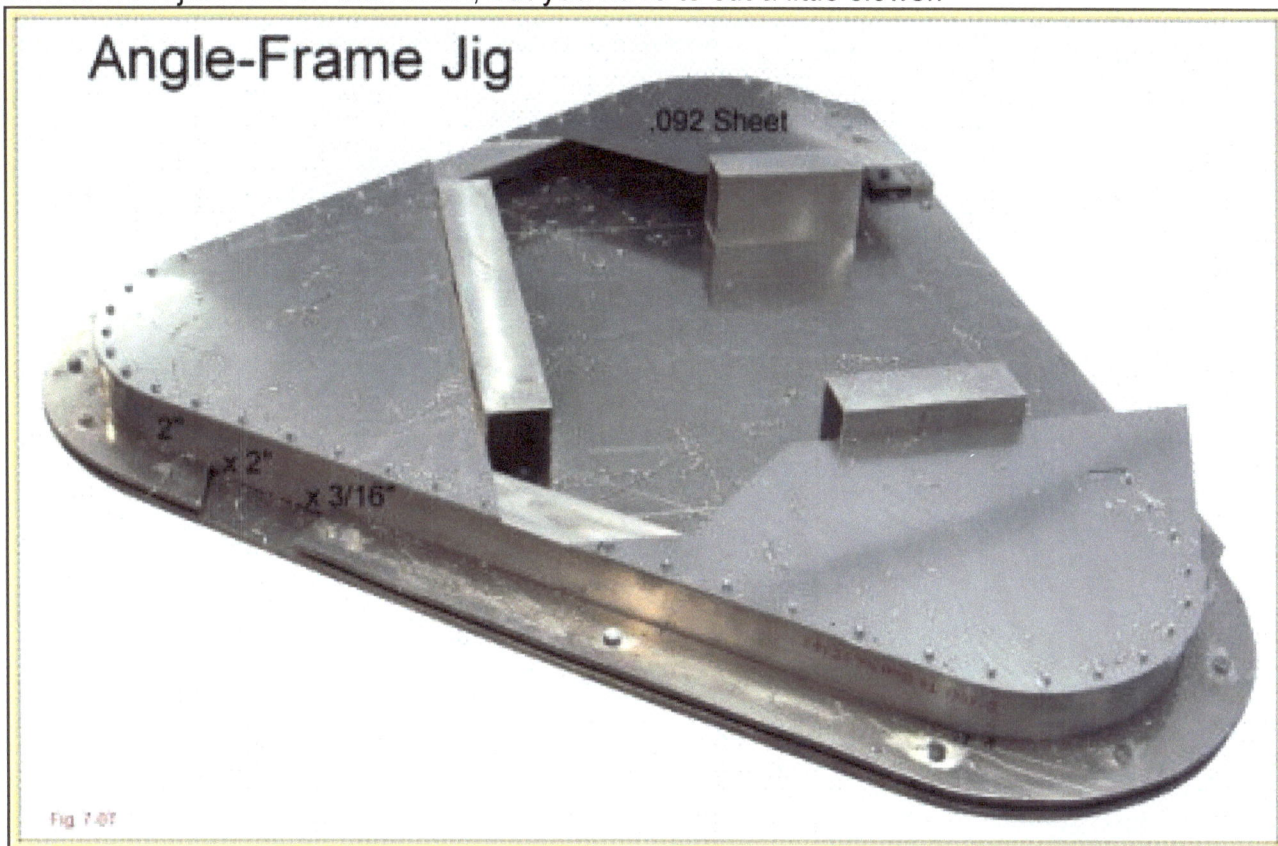

Bending aluminum angle is a simple procedure involving more cutting than you're probably used to and a forming jig would be necessary for radius curves but square corners are easy enough without a jig and you can bend them or miter them.

The angle-frames below for attaching the back plates above are the most time-consuming operation in our process: 45 minutes. The water-jet can do them, but in this case it's not cost effective and we do them ourselves. They start out as 2" x 2" x 3/16" aluminum angle and get jig-sawed into the origami-looking thing shown below. The 90 degree corner shown was gratuitous but relates in that it, too, is a means to fasten plate to a backing framework. They both involve wrapping aluminum angle around corners without welding. They both preserve the perfect flatness of a new plate. You can laminate and use your same carbide router bits to trim it. Using router bits you can fence-guide whatever edge profile you want on an un-backed plate.

Aluminum can be made to look very smooth and shiny by further buffing with the fine grit Maroon Wheel and then hand-rubbed with a Scotch-brite pad. Unfortunately, aluminum oxidizes over time and looses its luster so it's not recommended for any kind of exposed surface where it would be subjected to wear or chemical contaminants, including water. Viewed as a substrate, however, it's ideal for the application of paint, powder coating, and plastic, ceramic, or wood veneers. Its main advantage over wood for such purposes is that similar design functions can be achieved with far less bulk.

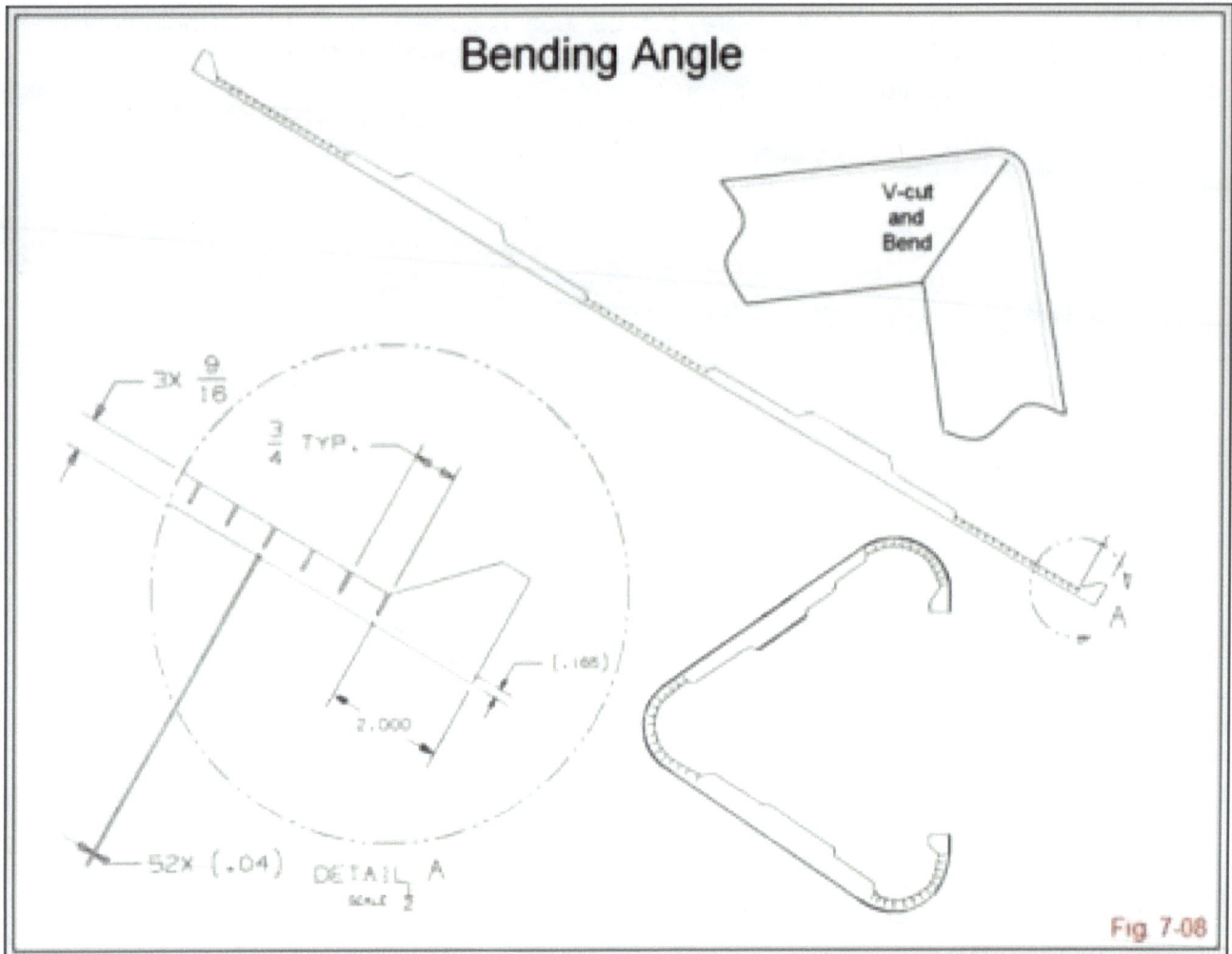

Bending Angle

V-cut and Bend

3X 9/16

3/4 TYP.

(.168)

2.000

52X (.04)   DETAIL A
scale 2

DETAIL A

Fig. 7-08

For the joining of bigger and/or thicker parts you'll want to use machine-threaded bolts and screws but you may or may not want to have bolt-heads protruding, depending mainly on whether you want the option of taking it apart again. One way is to counter-sink and use flat-head screws; the other is to grind the heads off. Counter-sunk joints are best executed while the two pieces remain clamped throughout the process, just as they are when making sheet-metal-screw joints. While the lay-out and clamping operations contain a margin for error, the drilling and tapping thereof do not. It might seem obvious to you, but it wasn't to me. I kept trying to take shortcuts and cross-threaded more than my share of critical joints. Done properly, you can install a 1/4-20 blind-hole tap in ten minutes, with a flat-head screw flush with the top surface. I use four drills, pre-loaded with: 1/8" pilot drill, 13/64" pre-tap drill, 1/4" drill (stopped to slip through only the first layer), and a 5/8" counter-sink bit. It takes about a minute to do the drill operations and maybe five minutes to tap the threads. I spend the other four minutes doing the boogie-woogie with my shop vac.

I use tie-down straps extensively for clamping operations but I cut the steel loops off and run the free end directly into the ratchet and fold it around the ratchet directly. Construction adhesive sandwiched between the folds has never failed for me yet. Clamp the sandwich in your vise over night and it's good to go. I use strap clamps to wrangle the angle-frames into my Titan jigs. They're great for wrapping any kind of straight things around any kind of round things to hold them in place while I'm inserting invisible screws. In conjunction with c-clamps they can be used to hold anything anywhere for as long as it takes. Cheap as they are I'll screw right through the belt and rip it out later if the operation is important. If you can bend a strip of 1/8" plate around a form, you can bend another strip right on top of it, thus creating the foundation of a bent ply; just as you'd do with wood. This is how the new prototype saw frames can be made by industrious do-it-yourself bandsaw designers. The finished product actually comes out cleaner and stronger than it would have been if you'd bent or welded a thicker plate. It's good to know that there are still some things that can't be done better by machines. As any woodworker already knows, there's no substitute for craftsmanship. This is just one more example.

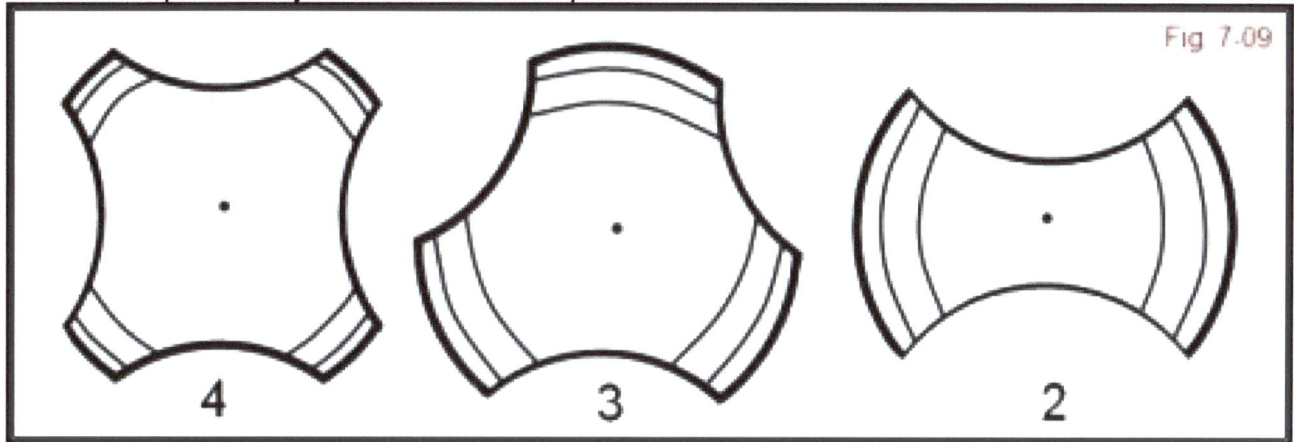

Fig 7.09

You'd be really surprised how much heat is generated by tapping threads. When you're tapping threads right after drilling the holes you have to take care not to lock up the tap in the middle of a tapping operation. This can be very aggravating and costly, if you break the tap off inside your precious part. Spray the thread-hole and dip the tap frequently to keep it cool and wet. You'll learn soon enough not to force the tap but if your parts are securely clamped you'll be able to back up a half turn for every two turns in. Doing that keeps your filings from growing too large to drop out of the flutes. For blind holes you'll have to flush the filings out with your spray bottle (although, when I can, I drill in far enough that it's not necessary). You can also clean out blind holes with your shop vac. There are days when my production schedule calls for the drilling and tapping of, literally, hundreds of holes so I've had plenty of practice at this and do the entire tapping operation by feel; looking out the window, sky-larking as it were. I wrap duct tape around the handles of my tap wrench to avoid blisters; nice.

I make my own blade guides here and each one has three 1/4-20 through-hole taps and a 5/16-16 through-hole tap. Being small parts, I do them in a bench vise next to a bowl of water, and rinse the parts after every operation. The tapping, as I said, goes very fast but the cutting is as difficult as anything I can think of. These brackets are made from 2" x 8" x 2" channel and have to be cut with a jig saw. My machinist can, and sometimes does, make them for me but it's expensive to have them made by machine too, because he has to mill the entire cavity where I cut it out in one chunk. The cheap trick to this kind of part making is to drill the corner turning points 5/16" ; just enough to insert a 318B jig saw blade. It's all straight cuts from there except you have to "rock" the jig saw over the front top edge and avoid banging the tip of the saw-blade into the lower channel-leg while doing so. As usual, it's a clamping problem and the trick is to support the part on a strong bench vise, using one hand to control the saw while the other is shooting water as fast as my fat little fingers can convulse around the trigger. It's a sloppy looking piece after the sawing operation but cleans up nicely with about fifteen minutes and two good files. I erase the vise tracks with a black abrasive disc and file the edges last.

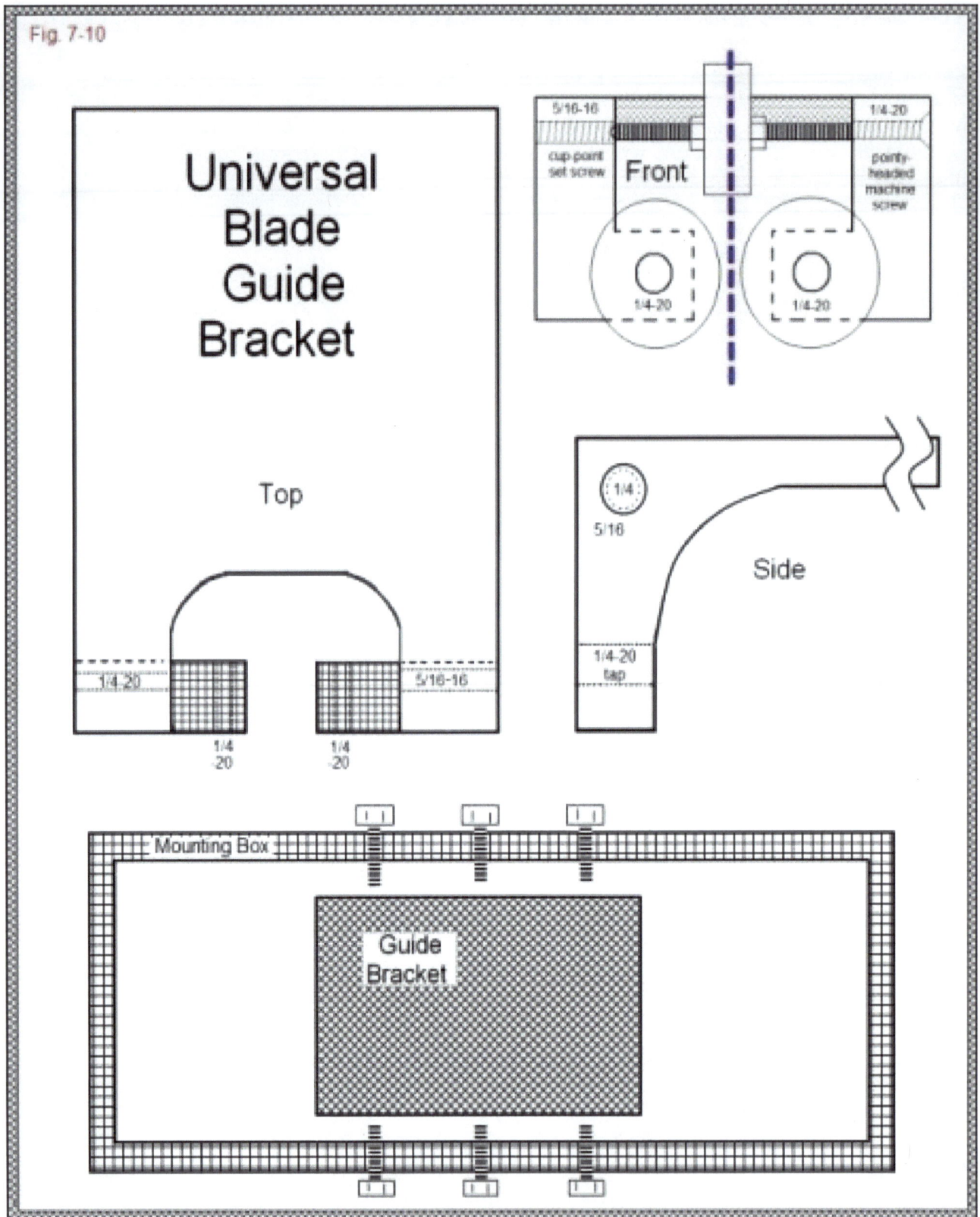

Fig. 7-10

Universal Blade Guide Bracket

Top

Front

Side

5/16-16 cup-point set screw

1/4-20 pointy-headed machine screw

1/4-20   1/4-20

1/4
5/16

1/4-20   5/16-16

1/4-20 tap

1/4 -20   1/4 -20

Mounting Box

Guide Bracket

I've had to bend sheet aluminum on occasion and found that some ways are better than others. As with most of the other suggestions contained in this chapter, it's not necessarily the fastest way nor the easiest way but perhaps the cheapest way to get the most quality with the least tooling for metal-working woodworkers. Aluminum sheet can be bent into nice, straight, neat right angles along well defined lines by sandwiching the bend between two pieces of square tubing and clamping as shown above. The entire sandwich must then be clamped to a solid bench whereupon the sheet metal wizard (that being you) will force the free end downward while tapping repeatedly along the edge to be bent. This requires much tapping and it must be worked back and fort, yielding a little more bend with each successive pass. The hammer blows need to be as close to the line as you can fit them. Such bends won't be as pretty as a brake would produce, but close enough you can conceal the "hammered" look under the powder coat or polish the dips out with a coarse abrasive disc. If the sheet is thick and the bend is long you can drill a line of 1/8" holes to make life easier. The closer the holes, the easier it will bend. *How* easy depends on how close you space the holes together. I scribe a line as deep as possible in the aluminum, then center-punch inside the scribe line groove for accuracy. Done neatly, the resulting "bead work" on the edge looks decorative. There are many origami tricks you can do in this manner that can't be done with a brake anyway, so it's a cool technique to learn in any event.

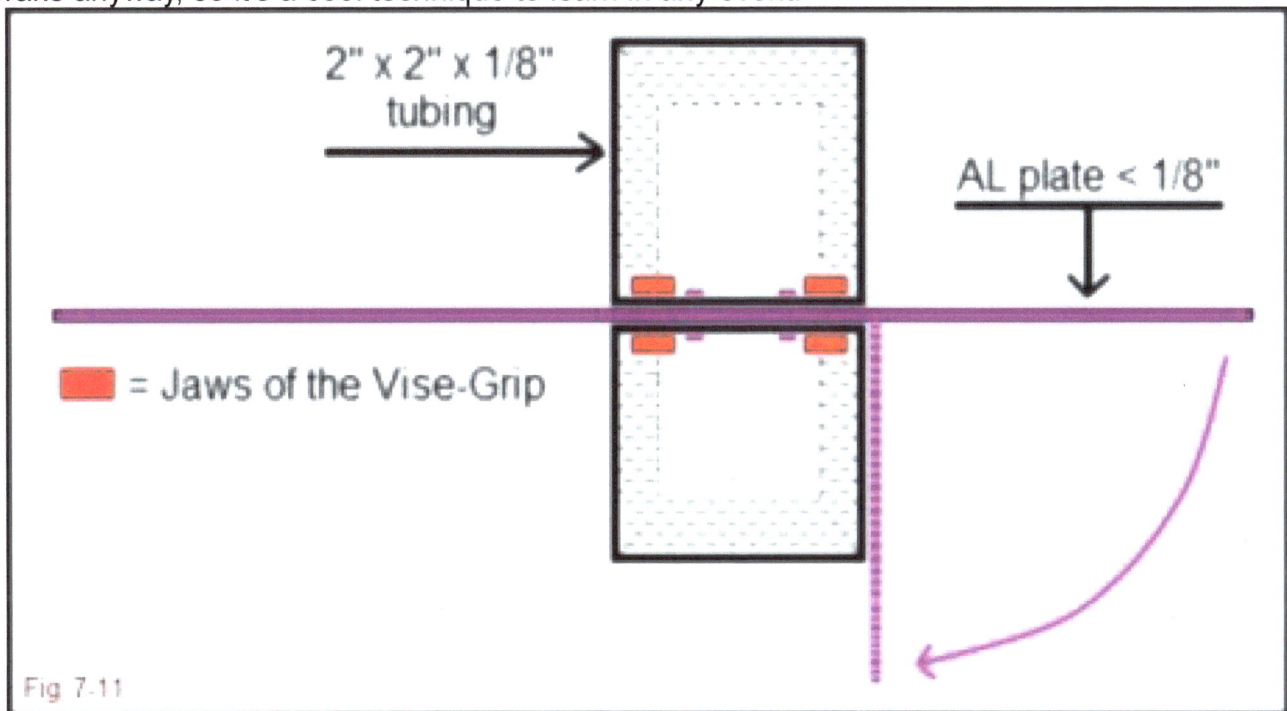

2" x 2" x 1/8" tubing

AL plate < 1/8"

= Jaws of the Vise-Grip

Fig 7-11

Nothing exotic to see here; just a basic, simple table; maybe lighter, stronger, and simpler than one constructed of wood, but it's basically four legs and a top.. There's no end to the possible variations of this same basic theme. Finish it nicely: it'll be around for a long, long time.

One such variation might be the three-legged concept. Wood doesn't lend itself easily to off-set posts and columns like the stands below demonstrate, but aluminum does. The saw stand below is functionally a three-legged table with an off-set column. It is extremely simple and easy to build, using the minimum amount of material, while providing a hundred or more pounds of support without obstructing "leg room" or, in this case, a Titan saw. The legs are 2.5" x 24" x 1/4" and are simply through-bolted to the 2" x 2" x 34" square tubing. The tubing was cut at 7.5 degrees off perpendicular at both ends using a v-cut jig that clamps nicely into my magic chop saw to get the rearward slope of the post. The legs are bolted perpendicular to the post but raised 2" to create, in effect, a three-point leg system which levels the top of the post, which in turn supports the horse-shoe looking plate that holds the saw with a blade in its center. Two 1/4-20 flat head machine screws are threaded, at an angle, into a 3/8" tubing end-plug to hold the top plate down. They're cheaply made but so effective. Future variations on the same principle should be seriously considered for movable breakfast nooks.

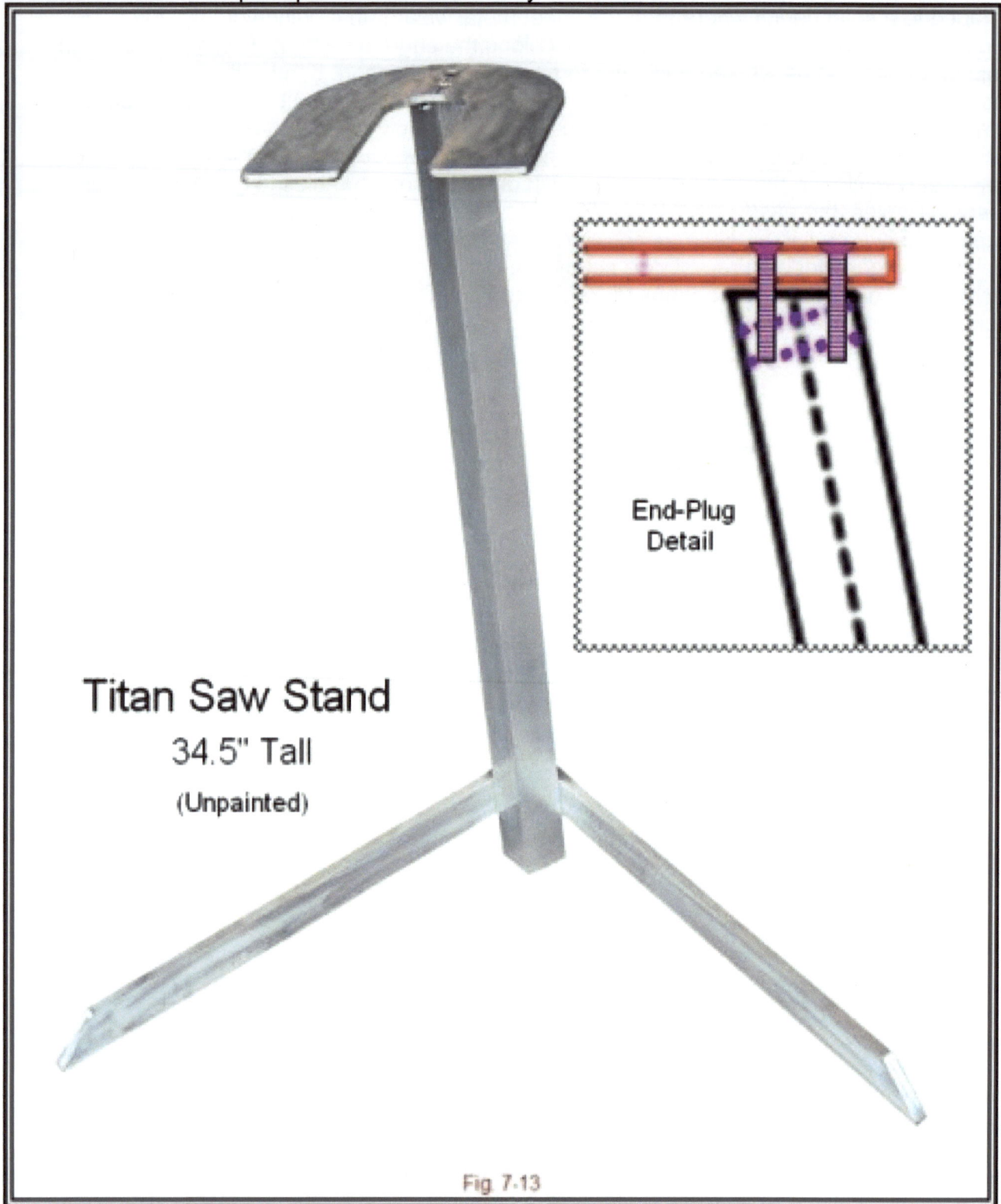

End-Plug Detail

Titan Saw Stand
34.5" Tall
(Unpainted)

Fig. 7-13

Another variation on the off-set column concept is used here to mount short timbers for portable saw cuts. We didn't always have timbers long enough to lay on saw horses with enough projecting wood to maneuver the saw so we cobbled this clamp-to stand out of scrap on hand. Since then we've used it as our sole saw-horse and have clamped some pretty long timbers on it; it's very, very sturdy.

The base is made from a single piece of 2" x 4" x 11' long tubing with v-section cutouts (made with a sawz-all) to make the two 90* bends. Inside the bends are two 4" lengths of 3/16" x 2" x 2" angle, secured with #8 hex head sheet metal screws. The corner braces are 3/8" plate and are secured with 3/4" long #8 sheet metal screws. The spine is 2" x 4" x 32" and is secured (loosely) to the base with four 3/8" through-bolts. The table-top platform is flat-head machine-screwed to 1/4-20 tapped threads in a 1.71875" x 3.71875" x 1/2" end-plug at the top of the column. It was immediately obvious that the top would need more support so two miter-cut 1/2" x 3/4" bars were blind-hole threaded at the ends for 1/4-20 flat-head screws. I don't know how many pounds it could ultimately support but we've hung some huge timbers four feet out over the top platform and saw it flexing some; but it seems the more weight we pile on, the steadier it gets; not bad for a jig you can pick up with one hand a throw it around the shop. It was made in two or three hours and has lasted five years now, totally replacing our need for space-eating saw-horses. Clamping plywood to it transforms it into a portable utility bench with a very small footprint.

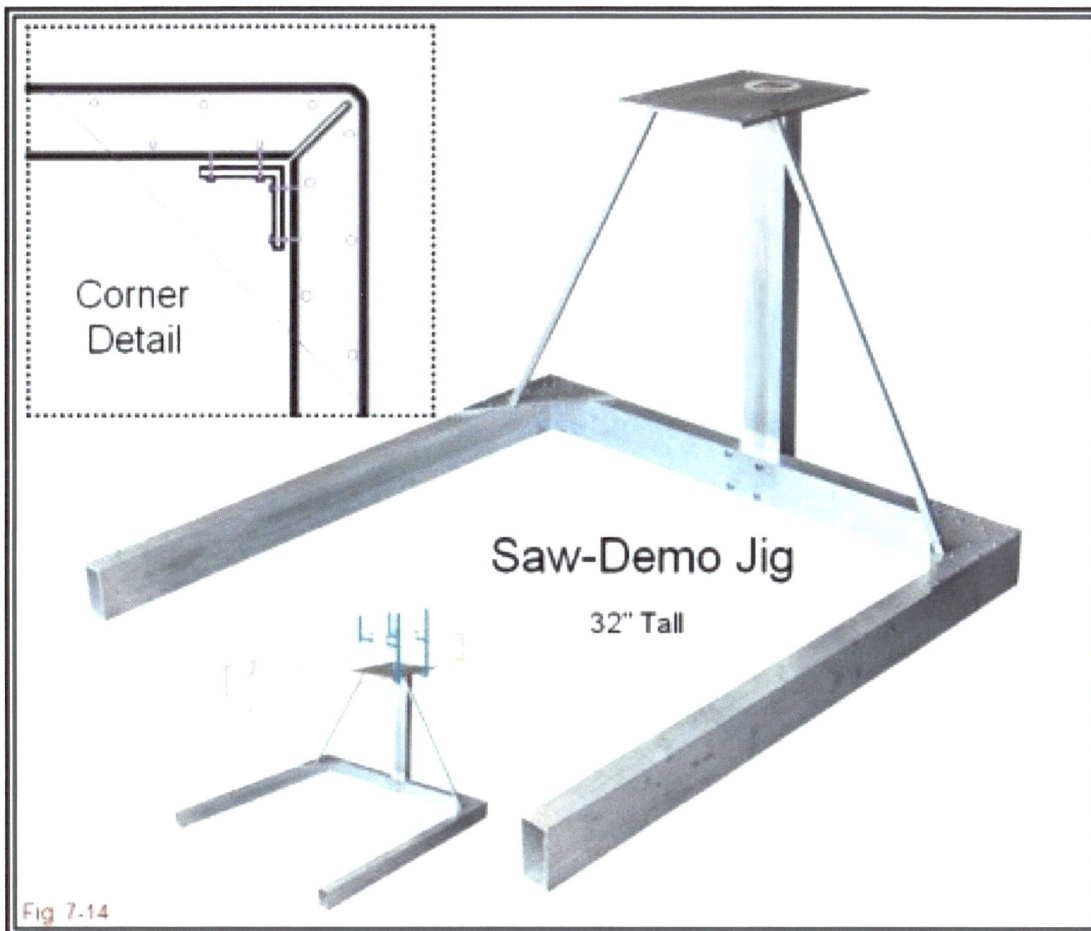

Corner Detail

Saw-Demo Jig

32" Tall

Fig 7-14

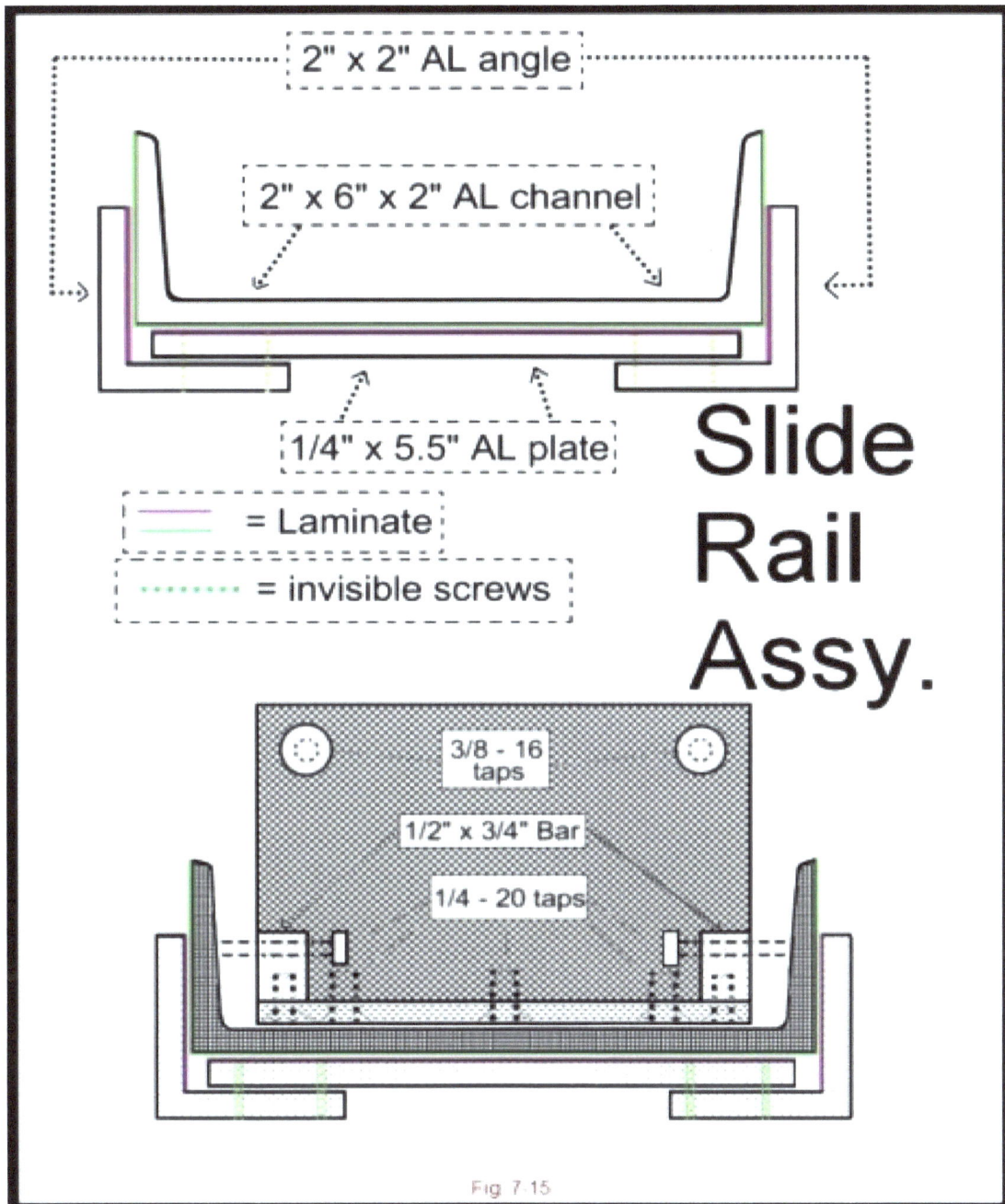

2" x 2" AL angle

2" x 6" x 2" AL channel

1/4" x 5.5" AL plate

—— = Laminate

·········· = invisible screws

Slide Rail Assy.

3/8 - 16 taps

1/2" x 3/4" Bar

1/4 - 20 taps

Fig. 7.15

If you remember the log-feed rail system from Chapter Five you'll recognize the drawing above as a cross-section of the inner rail, outer slide, and the three-point log clamp which locks into the inner rail adjustably. There's probably a million other ways to do it so I won't go into detail here except to say the outer slide assembly has to fit precisely in order for it to work. There's a lot of error in an 8' length of aluminum extrusion and you have to hold your tolerances to a single thickness of Wilsonart laminate, or its generic equivalent, throughout. I packed it all into a compact sandwich with clamps every 12" . I drilled into the inner rail just far enough to break through the 1/4" plate and didn't screw it down completely until after I'd taken the inner rail out; the heads and tails were then ground flush . The bottom laminate therefore, applied last, covers the assembly drill tracks;. The laminate application was no picnic and by the time I got that part of it mastered, I was ready for anything.

The clamp is relatively straight forward cross-cuts with lots of drilling and tapping. I used a belt sander to square it all up. Use good hard thumb screws and grind the ends to a fine point so they'll bite into the inner rail without too much pressure; too dull and the rail will spread under outward pressure: think PSI. Generally speaking, one of the coolest things about aluminum is its softness relative to steel. Bite a steel screw into aluminum and it stays **clamped,** baby! I won't say these log rails are an easy project but they are do-able by anyone with the skills of an experienced cabinet-maker using barely more than basic woodworking tools. Though I supplied some numbers, I leave the detailed dimensions to the reader to work out for himself because there too many possible ways to skin this cat and there is plenty of room to improvise around material acquisition problems and personal design considerations. I used mostly scrap for the clamp parts. Since the home-made clamps I used to attach this feed-rail to my saw are going to be so different from your application I omitted those drawings but by now I think you've got the idea.

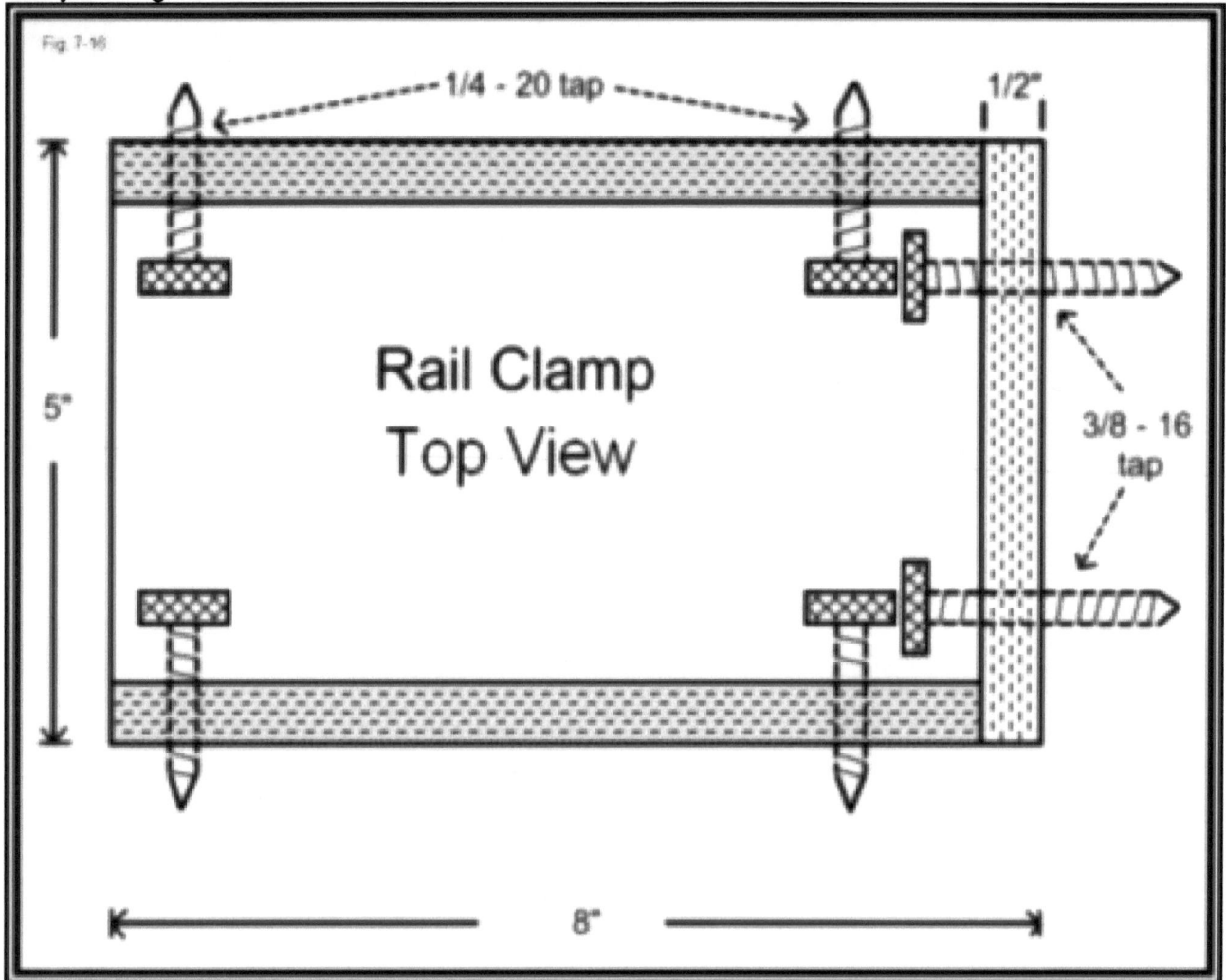

Fig. 7-16

Rail Clamp
Top View

1/4 - 20 tap

1/2"

3/8 - 16 tap

5"

8"

And while we're into mechanical drawing:

SliceMiester TT Assy.

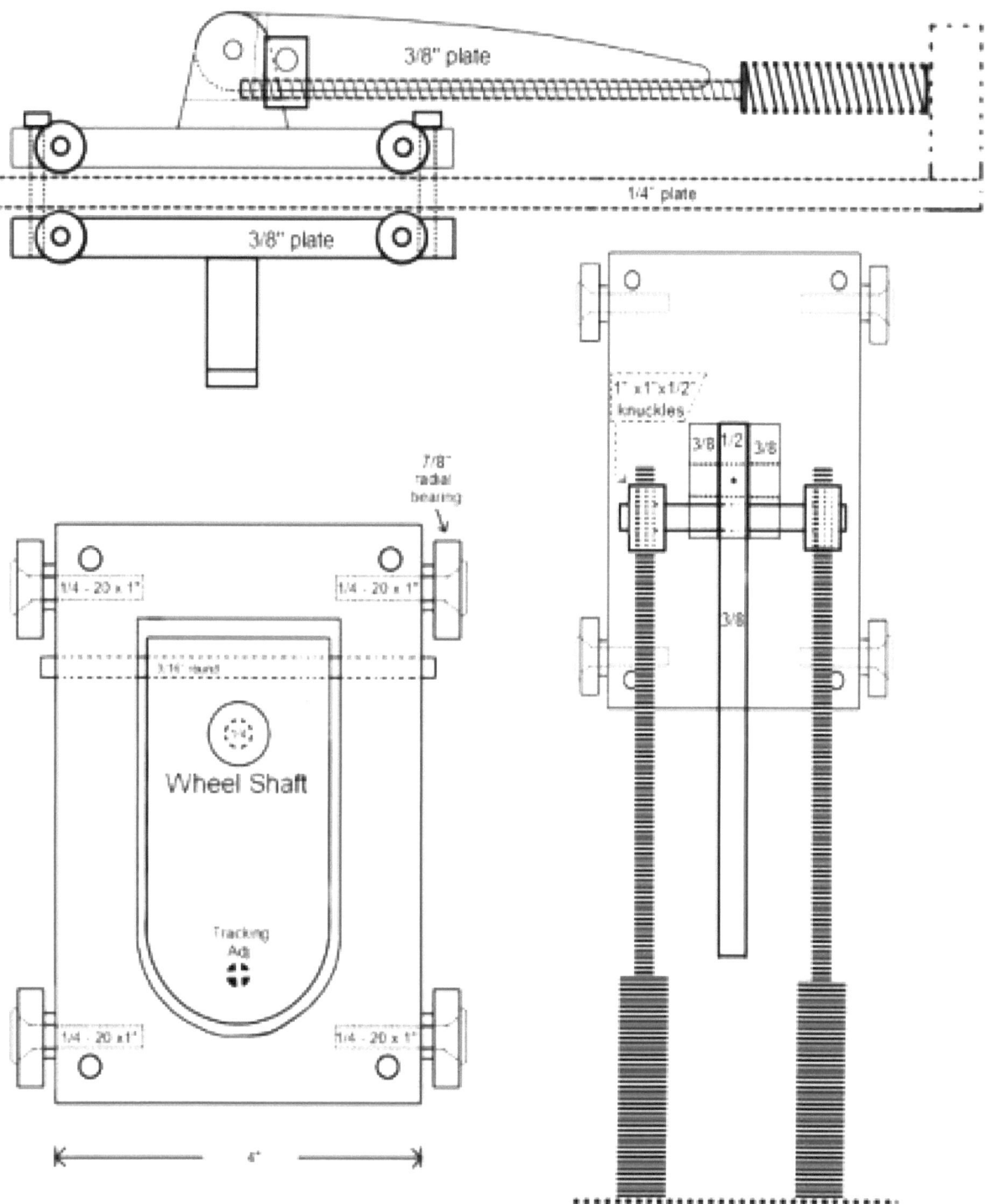

3/8" plate

1/4" plate

3/8" plate

1" x1"x1/2"
knuckles

3/8 | 1/2 | 3/8

7/8"
radial
bearing

1/4 - 20 x 1"

1/4 - 20 x 1"

3/8

3/16" round

Wheel Shaft

Tracking
Adj

1/4 - 20 x1"

1/4 - 20 x 1"

4"

3/8

Fig. 7-17

I'll throw this out not knowing if it makes sense to anyone. It's a complete tracking and tensioning assembly that you could install on any bandsaw if you can make the necessary alterations to its frame. The central point is to somehow replace your TT (tension and tracking) assembly's slide component for one that rolls on bearings. Using the old "bearing on a flat head" trick wherever it might fit in your machine would be worth doing if you can. You might have to cut your entire backplate out from behind your TT wheel; but if it's already trashed, -on an old machine...........................?

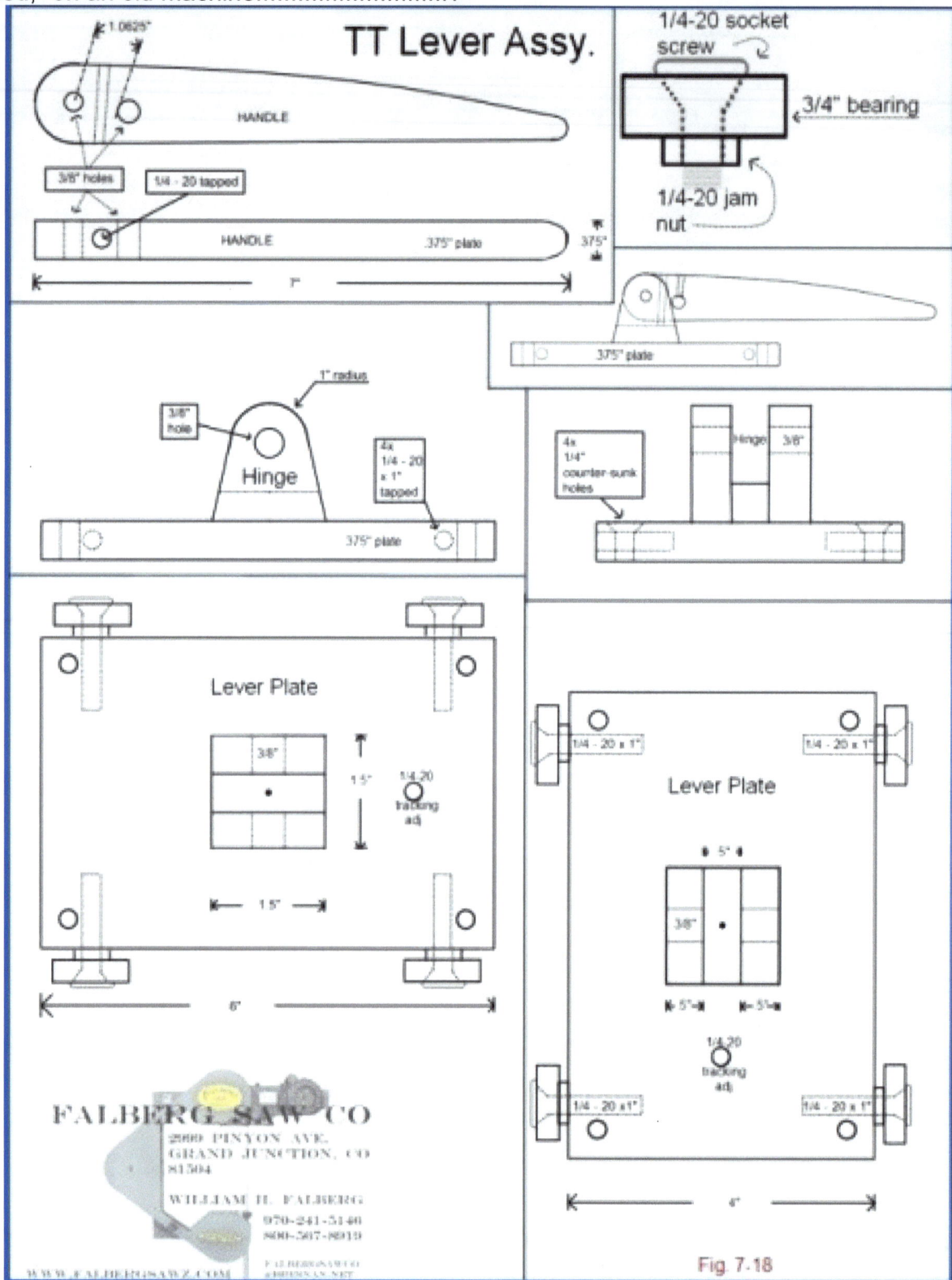

TT Lever Assy.

Fig. 7-18

FALBERG SAW CO
2660 PINYON AVE.
GRAND JUNCTION, CO
81504

WILLIAM H. FALBERG
970-241-5146
800-567-8919

WWW.FALBERGSAWZ.COM

And now for something completely different. Up until now this whole book has been about work. Work, work, work, work! And, frankly, it's been kind of boring. You have to kick up your creative heels once in a while and experiment with design concepts outside your normal realm just to satisfy your curiosity, if nothing else. The following is one such that I did six years ago, when I needed an entertainment center, didn't like the ones I saw, and couldn't afford one anyway. What I did have was some aluminum scrap and time on my hands.

Fig 7-19

Entertainment Center

6.5 Feet Tall

Wire Ties

The idea was to eliminate everything that didn't relate directly to supporting a piece of audio or video equipment and hiding the wires behind the skeleton of the frame with leaf-shaped ties of thin aluminum sheet. I had a half-sheet of 24 gauge that I didn't know what to do with and some odd shapes of scrap 1/4" plate. Back then I wasn't familiar with invisible welds and the joints are somewhat "clunky". I used flat head screws extensively. The framework itself needed further re-enforcement at the base; but that's how we learn, right?

Unseen in that tangle of leaf/ties in back is a 2" x 2" x5' square tube attached to the lower shelf with 2" x 2" x 3' angle, which angle was also used to attach the legs. The other vertical supports are all just scrap pieces that came out looking more like a bush, or a tree, so I thought the leaves were a logical theme for wire ties. Taste is a funny thing however, and what looks perfectly natural in my living room, next to the drums, is probably going to clash with most interior designers' idea of "elegance". I'll do better next time and I'm sure you'll do better the first time.

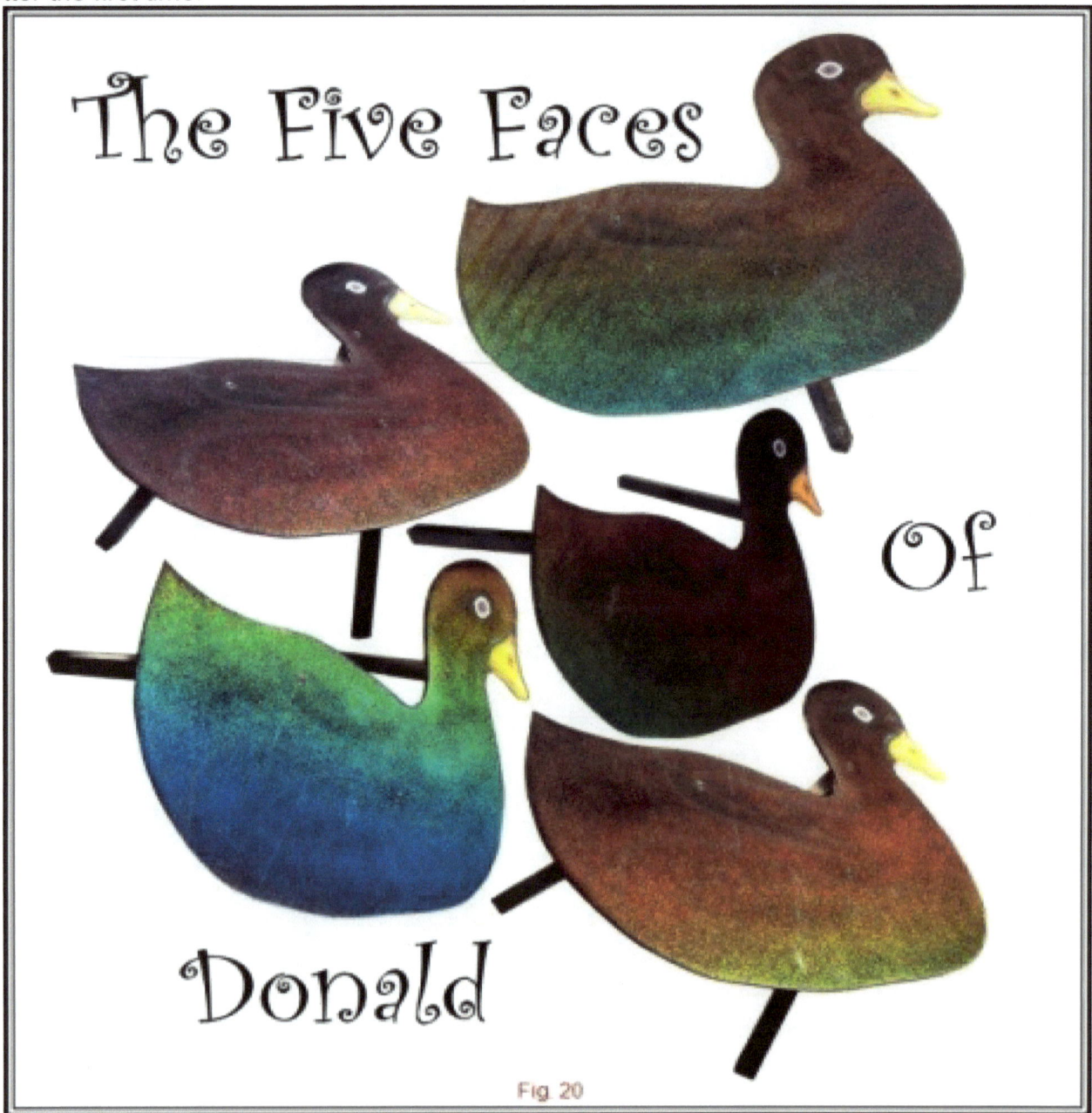

The Five Faces Of Donald

Fig. 20

Hard to believe it's the same duck, isn't it? The only thing I had to do to make the "Donald" table was round off the bill a little bit. It was a 1/4" plate scrap that just came out that way and the voices in my head made me do it. I had just enough 3/4" bar (that I didn't know what else I'd do with) to make three legs so I cut them at an angle and screwed them on. Having nothing more invested I took it to my powder coaters, Connie and Randy, of C&R Custom Powder Coating (970-874-7664) and let them decide how to finish it. I left it totally up to them to turn it into a duck-looking thing so they powder coated it with a chatoyant new powder of theirs called "City Lights". They then passed it on to another customer of theirs, Kevin Peebler, also of Delta, Colorado, who did the custom air-brush work on the bill and the eye. This was done before the entertainment center and I was so impressed with the way "City Lights" looked in the sun that I had the entertainment center painted with "City Lights" also.

Unfortunately the entertainment center never spent much time in the sun and, looking mostly black with speckles, the chatoyance is mostly wasted. It's still very cool-looking. I've often thought what fun it would be to just make aluminum furniture for a living. It gives me great satisfaction to create things I know will last forever in any environment, indoors or out. Aluminum's longevity aspect ought to motivate one to do do better design work but in my case such projects are a distraction and I don't spend much time on them. Some day I'd like to dedicate serious time to exploring the full potential of aluminum in a mixed media format for in-and-out-door furniture designs.

cd storage

Fig. 7-21

The CD storage rack above is more the child of necessity than any attempt to create serious furniture. Once again, it's made from scrap and its sole function is to store CD's safely. I gave each piece a once-over with the Maroon Wheel before assembling them because I knew it might be a long time before it would get the screw-heads ground smooth or powder-coated. As it is, it looks right at home next to the antique barber chairs in my living room and it's almost filled now. You might recognize the shapes of Falberg saw platens on the side-panels. Waste not/want not, eh?

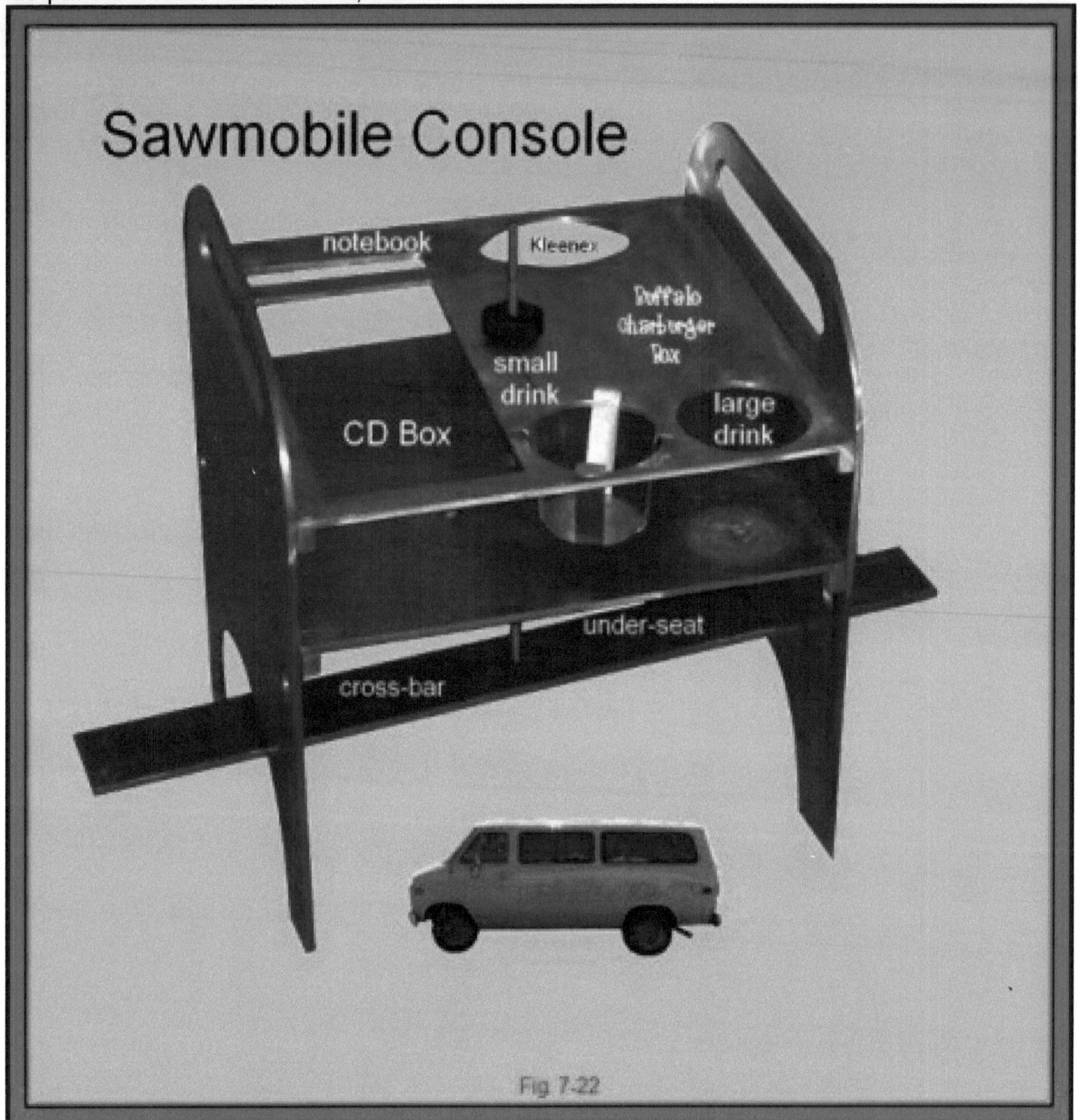

Sawmobile Console

notebook

Kleenex

Buffalo Charburger Box

small drink

CD Box

large drink

under-seat

cross-bar

Fig 7-22

The center console for the old saw-mobile is another example of quick-and-dirty utility designmanship. GM graciously provided the owners of these huge vehicles a tiny little utility tray on the "dog house" (that's "engine cover" to those of you not conversant in van-talk) about the size of a foot-long submarine sandwich. (And the drink goes where? What were they thinking?!) Granted, you need to remove the dog house for servicing; but you still need a place to put the CD's, coffee, road food, and kleenex on those day-long road trips. Why that wasn't provided for is a mystery to me. My solution made it removable by means of a crossbar between the captain's chair seats that tightened upward with a knurled knob on a threaded rod. The rest is obvious from the picture. This alone should be reason enough for any red-blooded American to learn some basic aluminum-working skills.

Mobile Base

Fig 7-23

These mobile bases are designed to lift, as well as move, heavy objects. In my case they were 500 pound antique barber chairs but the basic design would work as well for any massive iron machinery. If you look closely you'll see two flat iron straps hanging from two 2" x 2" lengths of steel angle. The knurled knobs you see on the mitered right side rail are for quick removal and allow that whole side to open. It is meant to enclose the heavy object by surrounding it and jacking it up off the floor with four vertical carriage bolts, seen here with vise-grips serving as wrench-handles. With this device one needs only to pry up one corner of said massive object and slip the flat bar under the object.

Intermediary shims are used to hold the object up while positioning the flat straps for lifting with the carriage bolts. It can, of course, be made to any dimensions. This was built to be somewhat universal, thus the extra angles-inside-angles angle. A dedicated lift of this design would simply have four outside frame members; one of which would be removable. The two vertical up-rights are just because the chairs swiveled when I tried to steer it and I'm too old to be bending down like that. I used 3/8" carriage bolts arbitrarily and they show no signs of strain but use your own judgment on this and other choices relating to scale. The corners are all miter-cut and the four base plates are 3/8" aluminum into which the wheel assemblies are tapped and hex-bolted. I used aluminum hex bar stock (scrap) to space the lift nuts above the angle; it's just faster to have the wrench there and free to turn when you're lifting. It's nice to not have to worry about tipping these things over.

Note that the weight is supported directly over the base of the carriage bolts, nowhere near the center of the flat bar. The holes in the flat bar should be just outside the base of the object to be lifted and, once lifted, not free to slide in any direction. I recommend a diagonal placement to prevent the object from slipping out of the bars' confines. The bases shown here were made to lift top-heavy barber chairs with circular bases; hence the wide wheel-base. Like putting on dirty underwear in the morning: orientation is everything.

Another cheap trick is to use a small square of duct tape to keep nuts from dropping out of vertically held socket wrenches. Likewise: a small square of duct tape across the top of an open-end wrench will hold nuts in place while fishing bolts through blind holes. If it's way down inside the tubing I tape the wrench to a longer stick. There's no problem that can't be solved with the proper application of duct tape.

I also cheated on my Bosch jig saw. I had my machinist custom-cut two base plates out of 1/4" tool steel. The factory, OEM, base-plates are too soft for the constant abuse they get cutting thick aluminum at the weird angles I often do. If you see your base plate starting to bend up at the front, like a ski tip, hammer it back down; tipping those saws forward when cutting adversely affects their feed rate. You don't need to go that far but if you find yourself doing enough aluminum work to justify the expense it's well spent. I use my Bosches so hard now that I'm starting to break the guts that make the thing go up and down, up and down, up and down, up and down.

In most cases when joining aluminum tubing at a mid-point of its length it is advantageous to bolt only the inside surfaces rather than bolt through their entire thickness. Bolting across the hollow core obviously raises the question of collapsing the tubing, and you never get a very tight joint. Doing so requires getting the nut and bolt inside while retaining the smaller bolt-hole at the junction. I haven't seen any significant weakening of structural integrity associated with the use of 3/8" or 1" access-holes adjacent to or in-line with the screw hole. You can slip a 1/4-20 socket head screw through a 3/8" hole but you'll need a 3/4" hole to fit a nut-holding socket wrench in the other side.

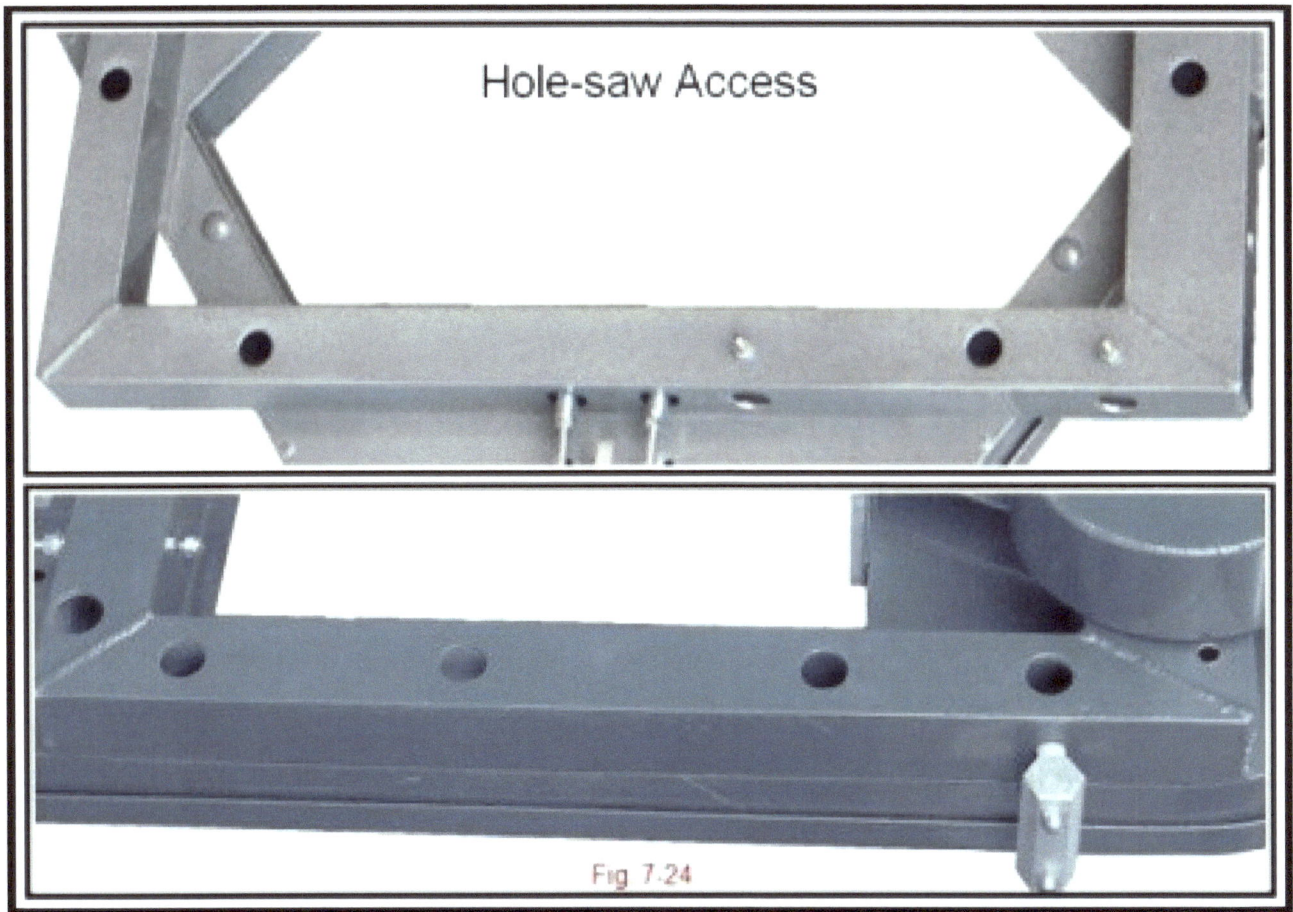

Hole-saw Access

Fig 7.24

I use hole-saws extensively in my business and found that hole-saws will drift all over the place if the drill in their mandrill's center is still cutting while the outer blade is cutting. I had my machinist make me a center-drill with only 1/2" of fluting. In other words: it was smooth-shafted after the saw teeth made contact with the surface. That effectively stopped the "wallowing" of my 1/4" center holes and the sawn-holes, thus drilled, immediately smoothed out; the process becoming much faster and easier as a result.

There. Now you're a master metalsmith. Go to work.

CHAPTER EIGHT
What You Can't Do Without A Bandsaw

Maybe, now that we're done fidgeting with our stupid saws, it would be appropriate to undertake some fun projects, the likes of which you don't normally see in bandsaw books. The only experiences I have to share come from butchering monster chunks of hard wood into usable chunks of hard wood. My favorite and certainly the most fun was the driftwood episode which I now call Pescado Mysterioso that has since become a legend on all four continents. The saga has been re-told so often that it's become almost unbelievable so I want to set the record straight and put the rumors to rest.

It started in the winter of 2004 when my sons and I were visiting family in Santa Barbara, where dolphins swim in-close to watch the tourists stroll the beach. It was a warm, sunny day and I thought the pelicans were acting strangely as if they were about to perch on our shoulders. The dolphins seemed to be watching us with particular interest, also; but we're used to being stared at and paid no mind. Otherwise, the beach seemed as normal as ever while my youngest and I cavorted from dead mollusk to the creepy things on the rocks in search of a distraction. Being artists, we never get to the beach early enough to find any collectible flotsam and you need wet-suits to reach the jetsam; but I comb for shells nonetheless, every time I go there. One never knows.............

I remember it all as if it were only yesterday. We walked West, toward UCSB from the Arroyo Burro Park looking as cool as any two guys can be for about three hundred yards to where the long slimy green things are abundant. I figured if there was anything worth picking up it would most likely be found in the most repulsive stretches where the sand flies are thickest. And sure enough; I was right! There before me was a genuine piece of wood that had drifted to shore and had miraculously waited for me to come along and pick it up. I picked it up. It was indeed a good solid real piece of driftwood. I was all excited; my son less so, but he's not as sophisticated as I and didn't fully appreciate the wonderfulness of it all. Whirling like a depraved dervish, kicking up everything that poked out of the sand, I scrutinized every anomaly in sight. Only those who have prospected for gold would understand the elation I felt at that moment.

It wasn't much longer before I found another, much larger, piece of the same vintage and texture as the first but, by its size, revealed its true arboreal nobility. It was too heavy to be anything other than an exotic hardwood of species unknown . It was about four foot long and shaped like a fish. That was my first impression and my son agreed; it was meant to be a fish in its natural state and displayed as is; unsullied by human hand. It did , in fact, remain untouched for the next two years while I thought about it, night and day, agonizing over what I should do with it. But I'm getting ahead of myself; we haven't got it to the truck yet.

I was taking one last look around to see if we missed anything when my eye was drawn a short distance off to a curious hump that didn't quite look like a rock. Curious minds, right? It didn't take much scraping and poking with my pocket knife to determine that the larger lump was identical in color and composition to the first but much, much bigger. We spent another hour kicking around the edges to see how long and wide it was. It boggled my mind! Yes, I was boggled; but only for an hour or two. My best guess, based on what I could dig up, indicated a tree trunk roughly three feet in diameter by thirty feet long. Pure, un-cut, solid, dense, hard, and precious exoticus arboristicus!! Wood porn of the most seductive nature; just there for the taking. Ay; and there's the rub, me matey! How to get it home. How to get it home.

Driving a track-loader down a public beach in California would probably get you sentenced to fifty years of hard labor at the wind farm and since I don't the know anyone with an ocean-going salvage tug all I could do was stand there and drool. I have no idea how to reclaim such a treasure but I still dream of taking it home some day. I think, if I knew for a fact that I only had a short time to live, I would drive out there in the dead of night with a tow truck and attempt to dig it out under the cover of darkness and winch it up onto a low-boy; so ardent is my desire to cut it up. It's still out there and I'll come for it some day. I located it on Google Maps just recently. It's still there: just before you run out of sandy beach, this side of the cliffs; up by the high-water line; that little shadow laying sideways to the left of the big rock. By the way, the surf was bitchin' when Google took that picture; check it out, dude!

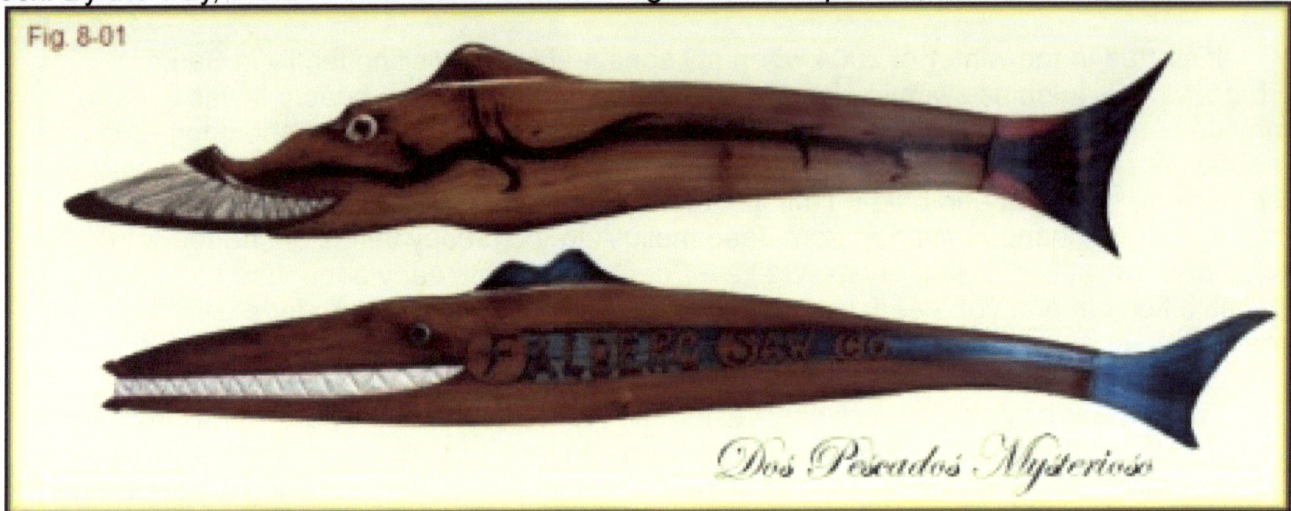

Fig. 8-01

Dos Pescados Mysteriosos

So anyway, there we were, three hundred yards down the beach with two hundred pounds of shark-looking driftwood and the smaller gecko-sized piece. We talked about it and I agreed he should take the big piece and I'd manage the little piece back to the truck. Nothing of further interest happened after that. We got home with our treasure. It sat for a couple years, waiting for the Corbel King to fully develop and my oldest son to get comfortable enough with its operation to start opening the shark for exploratory woodworking. Nobody knew yet what mysteries lie within the charcoal crust of its tumble-polished exterior and our curiosity was piqued. There was no way to know how deep the black discoloration went nor even if it was the result of burning but I knew it had to be from the Great Tree and at least 15,000 years old. I had done some antiquities research and learned from anonymous sources inside the Knights Templar that not only could it be from the Great Tree but that it could be the last known hiding place of the Holy Grail; as it was common practice in those times to hide priceless artifacts inside the knot-holes of big mysterious trees. This is all pure speculation, of course, but my Illuminati friends always wink when I say that so who knows.......................... The only thing we are sure of is that the sparks coming out of the kerf were no less than stupendous. They were so bright, in fact, that our cheap camera even picked them up on the video we shot of its opening. We're not prone to exaggeration here so suffice to say it was a moment that changed our lives in ways we cannot yet fully comprehend.

Zubin usually shoots our still photos but, since he was doing the cutting, all we got was the video that Amadeus shot. I had to edit a video clip extensively to get this image of that fateful moment when the Great Spark filled the screen:

We knew there was some powerful strong magic being unleashed by the crackling in the air and by the way the flowers in our shop were behaving. They normally just stand around ankle-high absorbing shop dust and exuding that elf-like aura and pleasant scent our shop is widely known for, but as soon as we opened that lost fragment from the world before time they came to life in ways we'd never imagined.

Our first tentative cut was intended merely to see how deep the black discoloration penetrated into the wood and we were somewhat disappointed to find it went very deep indeed. Furthermore, there were checks and cracks distributed evenly throughout that couldn't be cut away and would have to be filled. Our original plan to retain its fish-like shape in the round was revised for one that sought to find clearer grain definition at the center-line and making two fishes out of it, each having a "good" side. That, too, turned out to be a challenge as even the inner core was badly discolored and was rotted in so many places that even the original profile had to be completely cut away. I hadn't intended to carve any of it but digging out the rot with a pocket knife produced the same effect. The eyeballs became protuberant, the split down the center became greatly exaggerated, and the area around the teeth was so discolored that I had to paint them white. Both Pescados are finished on both sides, carving and all, but the details vary somewhat from side-to-side. You can't see it in the photos but all the hand-carved eyeballs are from the same knot and their polished concentric rings are visible through the white stain. I had to epoxy around them to keep the knots in place.

Zubin's exploratory cuts were done free-hand with a 1/2"-2TPI re-set blade at low enough tension to permit rough handling and arbitrary course corrections mid-way, but the cuts were remarkably straight and flat nonetheless. After two hours and three rip cuts we could see what needed to be done and the plan became what you see above; two big goofy-but-lovable fish.

Fig 8-03

As you can see by the depth of the splits and extensiveness of its discoloration, it wasn't going to be a "clean" carving in the classical sense. We knew the split down the center was complete but all the other little splits ran in every direction. You can see here why I had to ink-in all the little curlicues on the widest slab (El Pescado Grande) and the lesser ("Falberg Saw Co" sign) piece was not planed out flat. I was determined to get two big fish out of that wood, no matter what I had to do. If you look closely at the shop floor in the background you can see how the flowers grew to grotesque proportions in the short span of time between these rip cuts. It got so bad we couldn't even move around the shop, much less find our tools, so we had to defoliate the shop with a flame thrower the following day. I've since re-planted with daffodils; they're easier to walk through and don't eat the cords.

The next four cuts were done with a disposable sled: two of them all the way through, to make the two slabs. The last two cuts, through the center and ending where the split ended, were made to created a kerf into which 1/8" aluminum plate could be inserted. Thus re-enforced, it was ready for final shaping and finishing. I have to confess; I didn't create the Pescados. They were already there in the wood. I just cleaned them up and made them presentable for family gatherings. In the end we learned that you can take a fish out of the sea, but you can't take the ocean out of the fish.

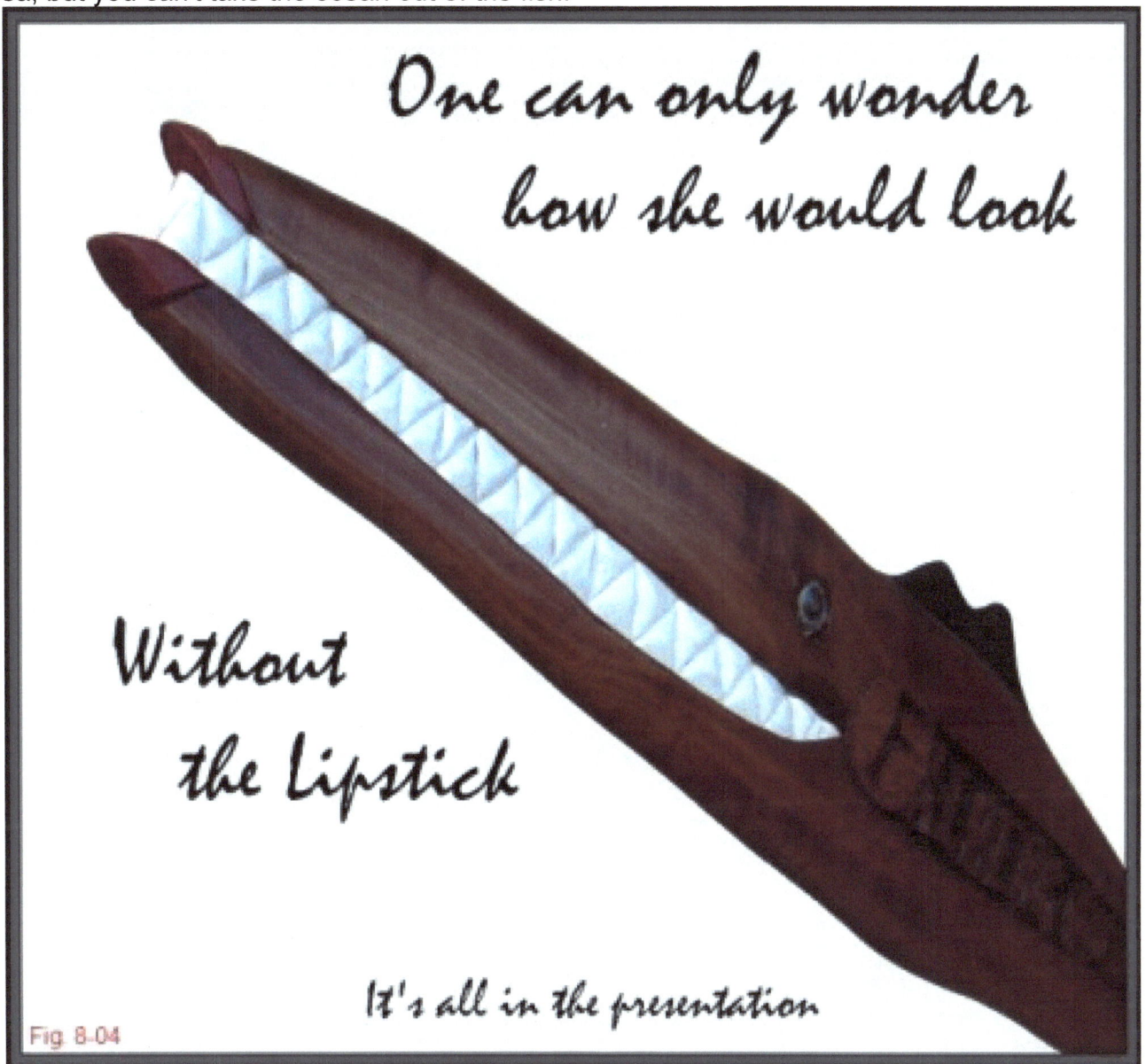

One can only wonder how she would look

Without the Lipstick

It's all in the presentation

Fig. 8-04

The aluminum plate had to be epoxied in before it was strong enough to carve, but thereafter it was quite stable and I used a 4.5" flap sander to shape most of it. The details were carved with chisels and hand-sanded. Finishing the aluminum plate was interesting because I tried a new (to me) technique with magic marker and polyurethane. I learned, unexpectedly, that magic marker colors will run when you paint over them with polyurethane; but (when you clean it all up again) that can work to your advantage if you pre-plan for it and *work* it while wet. You get very cool color-blending. With a little practice you soon learn how to *flow* the colors with swirls and in the end you've got a very bright, glossy, hard, metallic finish. I have not yet fully explored all the possibilities this technique portends for aluminum finishing, but I look forward to it. It looks like anodizing but with a pattern.

So bandsaws are good for making cool things out of found wood. For this reason alone you should think twice before settling for less than 16" resaw capacity.

They're also good for making money in the construction business. You can make beautifully sculpted corbels and rafter tails. Both require big throats, but rafter tails just about have to be portables with big throats to make them practical. That's what my saws are for and in case you haven't noticed; all the projects in this book feature big chunks of wood. All these projects were done with relatively low tension on a 1/2" blade. You couldn't see it here but during the free-hand cuts executed by my son on the Pescado project the blade was horribly deflected; at times by as much as three inches from its nominal path. Following scribed lines on a squared timber is quite a bit easier to do and it's amazing how accurately one can rip 12", 16", and 18" beams.

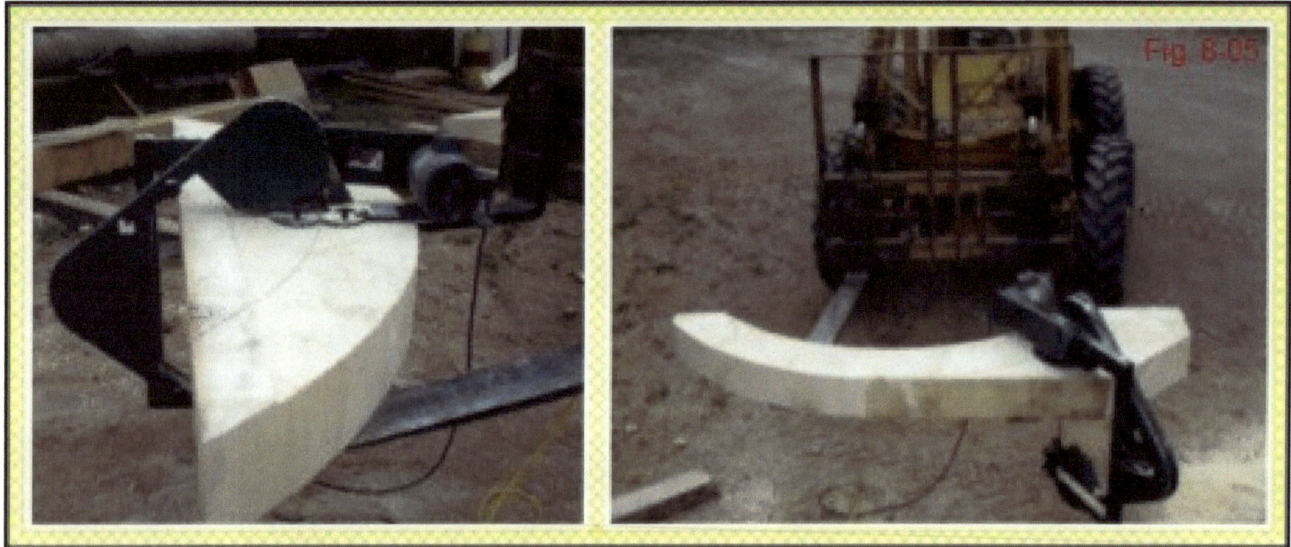
Fig. 6-05

Following scribed lines with either a portable or stationary bandsaw can be challenging if you're not aware of the fact that your hands are just another form of fence. Just as your blade needs to be on a plane parallel to a straight fence, it also needs to be parallel to your hands, but vise versa. While you can see and feel when your blade is parallel to a straight-line fence, you can't always see or feel such mis-alignment when cutting contours. I talked earlier about watching the sight-line from top to bottom wheel and using the top guides as a visual reference and it still applies, but working with big pieces introduces additional complications. You can't just let go of the piece and wait for the blade to self-correct. It won't. You have to force the workpiece, or the saw, back into line frequently. Such force should be applied directly perpendicular to the blade's path; and immediately. It doesn't do any good to *cut* your way back to the proper trajectory. It only prolongs the error and makes it worse. In most cases, if you've been watching carefully, you'll see such mis-alignment before the blade has had time to deflect in the kerf. The trick to staying on the line is to keep the teeth on the line. Slow down if necessary. It's still faster than sanding and grinding; or, worse yet, re-doing. Free-hand cutting is arduous work and commands respect as well as premium pay. For the operator, it's like *carving*, on steroids; I love it.

I had an Amish gentleman call about my portable band saws years ago and I described to him what they do and how they advanced the craft of timber framing by adding curves and sculpted details. His reply was; "Yeah. I heard you guys (out West) did a lot of *funny business* with your framing out there." The photo below is one such example of *funny business* by Jody Card; fisherman, timber framer, inventor, customer, and all-time Great American. Recent advancements in portable bandsaws have pushed the concept of *building-homes-as-if-they-were-large-pieces-of-furniture* ever closer to reality. The decorative elements of timber frames are even being used in stick-built homes as accents, for their artistic appeal.

Fig. 8-06

Can you imagine doing this with a chainsaw? Without a chainsaw? No. We'll see far more of this kind of work in the future and it will be done with bandsaws, but they will be much bigger bandsaws than the ones we use today: both portable and stationary.

There's an endlessly renewable source of precious hardwoods to be found in the tropical forests of Africa and South America and I foresee the woodworkers' demand eventually being met by savvy logging companies in cooperation with increasingly savvy governments. The logging industry of the past is undergoing the same sort of informational renaissance that the world's other trade practices are undergoing and I think we're already seeing signs of international co-operation in the import and export of precious hardwoods. Eventually even environmentalists will have to admit they don't know what they're doing. This is a pretty serious issue that woodworkers should all be aware of and I encourage everybody to investigate the matter of logging in Brazil, for one instance. The truth has a funny way of getting around and it's the truth that's going to make woodworking more fun than ever. You are hereby cordially invited to join in. All you need is a bandsaw.

Fig 8.03

These are some more of Jody's trusses. He had to stand them up and strap them to a low-riding platform to prevent trampling by marauding cape buffalo. As the effects of global warming become felt more widely throughout the Western states we'll no doubt see more of this and it's logical to assume that sturdier log and timber framed homes will increase in popularity. Although I've been accused of *photoshopping* my graphics in the past I can assure you that canaries really do piggy-back on cape buffalo to escape snakes. As a minor point of interest I call your attention to the size of those timbers. Can you imagine maneuvering those through a stationary vertical bandsaw?

There's probably a lot more I could say about the coolness of working big timber but I think the following pictures tell the story better than I could put in words. They all have two things in common: they're bigger than a bread box and when the off-cut drops off, watch your toes. The first two were done by my sons for demonstration videos; one finished, the other not. They were done in Douglas Fir and took about ten minutes. We usually hand sand them, if we sand them at all. Timber framers are sometimes asked to leave the finish as rough as possible to match the original saw-mill finish, that's even easier.

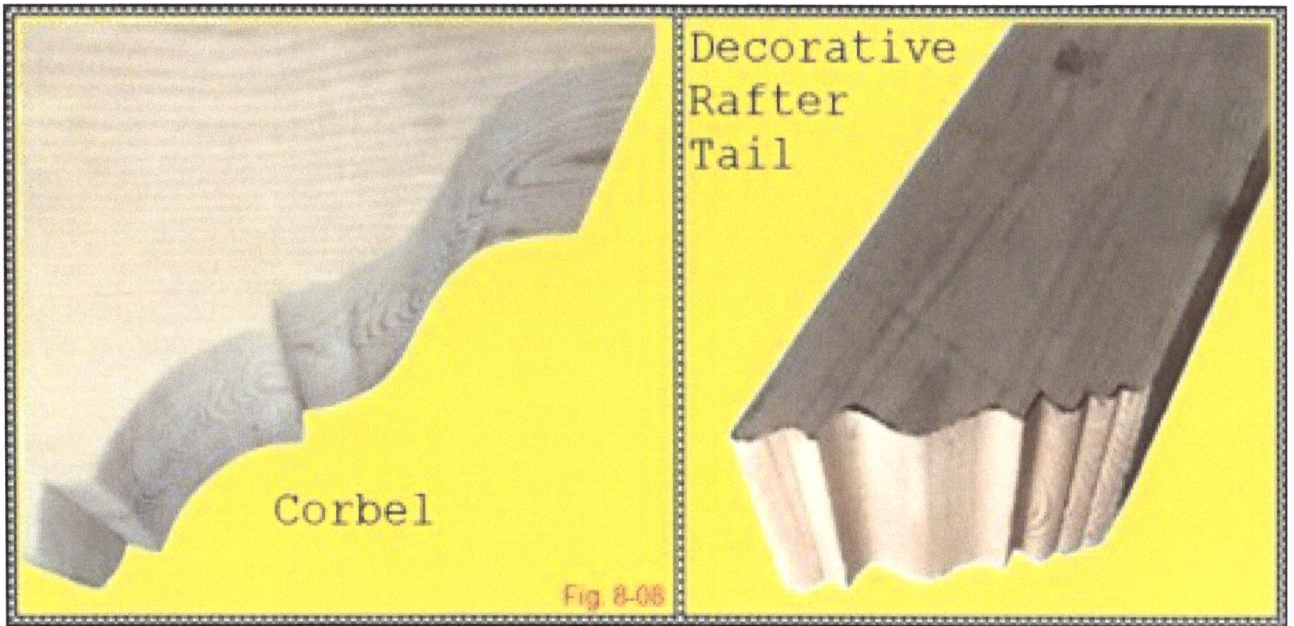

Corbel

Decorative
Rafter
Tail

Fig. 8-08

There are no end of architectural accents one can do with portable bandsaws but mantels, newel posts, and solid wood stairs come to mind. When one thinks of the band saw blade's off-set teeth as a power-rasp and runs the saw backwards against an exposed wood surface it becomes remarkably easy to achieve smooth, rounded contours. I've used mine to round square edges by manually rotating workpieces over the blade using the table for support. Just be sure to keep your fingers out of the way.

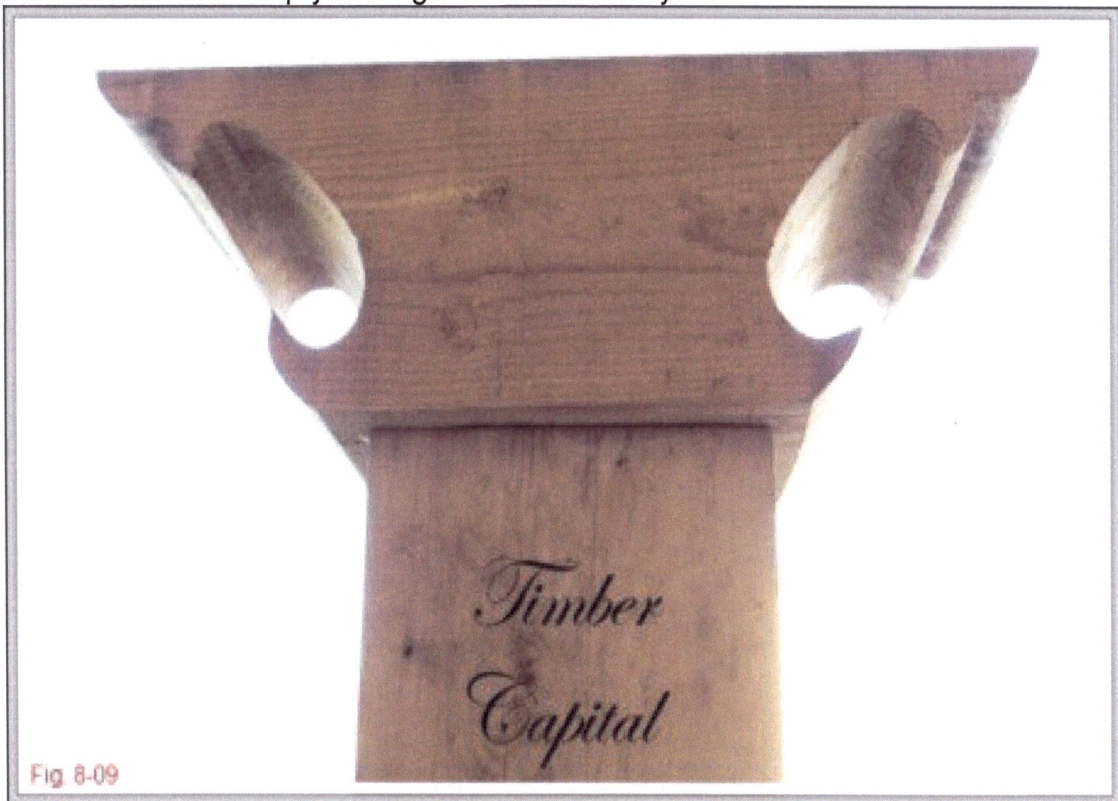

Fig 8-09

Timber Capital

The capital above exemplifies the advantages of a wide throat. It's 15" thick and the piece had to be turned over to attack the cross-cuts from opposite directions. Operations like that magnify the importance of keeping a straight up-and-down kerf because the blade has to be backed out repeatedly.

Unless the kerf is straight, backing out of a cut can be very difficult. Such operations also highlight the arduous nature of maneuvering the saw/workpiece manually to keep the blade between the wheels and not permitting deflection. You might get away with it going forward, but if the *back* of the blade *catches,* you've got a potential *un-tracking* event. Going back to Chapter Three, the importance of wheel crowning becomes more significant than ever. Doing the *back-up* in cutting the capital had my blade running wayyy forward of the guides but no blades were lost and I count my experience for that. It only takes a few *untracking events* to get the feel of how far you can go and it doesn't happen very often any more.

It's all about cutting solutions, of which tooth set and tension are key components. The *it* I'm referring to deals with dark, mysterious forces beyond our understanding or control. *It* is a black hole of ever-changing variables; *it* ebbs and flows with same elemental forces that govern the timeless order of the universe. *It* has many names, but I call *it* the cutting solution and weigh the parameters in accordance to their cosmic nature; invoking truth and beauty to materialize my woodworking dreams. Use the Force, Jody; use the Force! I digress.

There may never be much market demand for Bayer posts, but the technique to produce them coincides with a growing architectural trend towards more intricate timber frames featuring the flowing lines of arched trusses and the sculpted ends of exposed rafter tails. Timber frame homes, the last bastion of traditional construction, have become the ultimate expression of opulence. They have become more than just a frame work to support walls with a roof; they have even transcended the notion of being structural furniture; they have become works of art in their own right. Ironically, the idea that architecture was itself an art form and not merely a media upon which to hang paintings, was an idea that Herbert Bayer himself, espoused. In spite of this ideological serendipity, I have yet to see his "Totem" creation incorporated into a timber frame home. There are undoubtedly code issues involved with using them as structural elements but I remain hopeful they will someday be utilized decoratively, for they are very sound structurally and I think vigorous stress testing would prove that. Can you see an entire colonnade of them strung out along the entry to your place? It wouldn't be difficult at all.

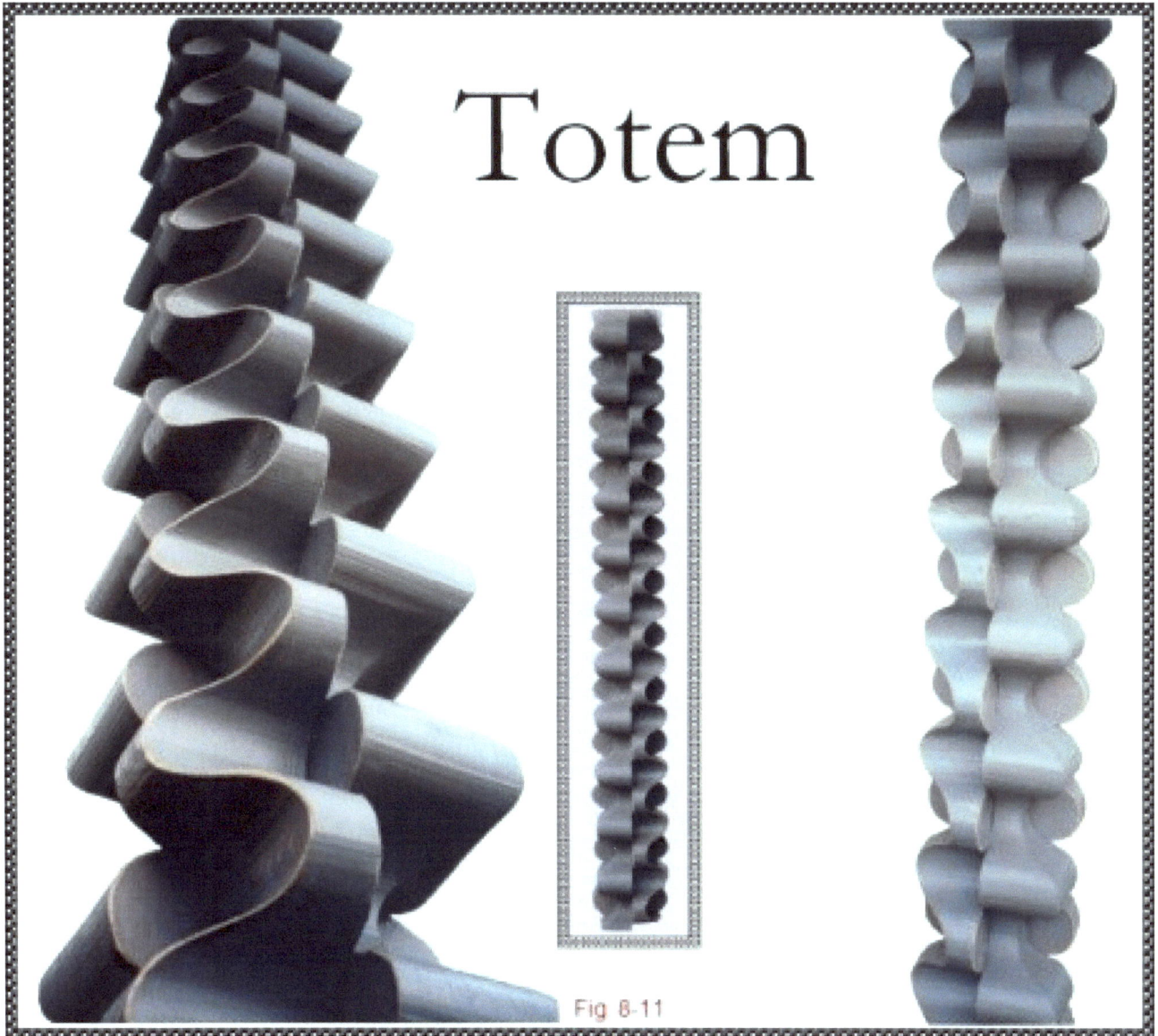

Totem

Fig 8-11

Imagine the effect you could create by exposing the rich, dark, heartwood of a tree with rounded surfaces reflecting off each other in such an intricate pattern. Used as supports for a gazebo it would be quite dramatic. Do you suppose there are are any craftsmen capable of such a feat?

Bayer Post Lay-Out
For 12" x 12" x 12' beam.

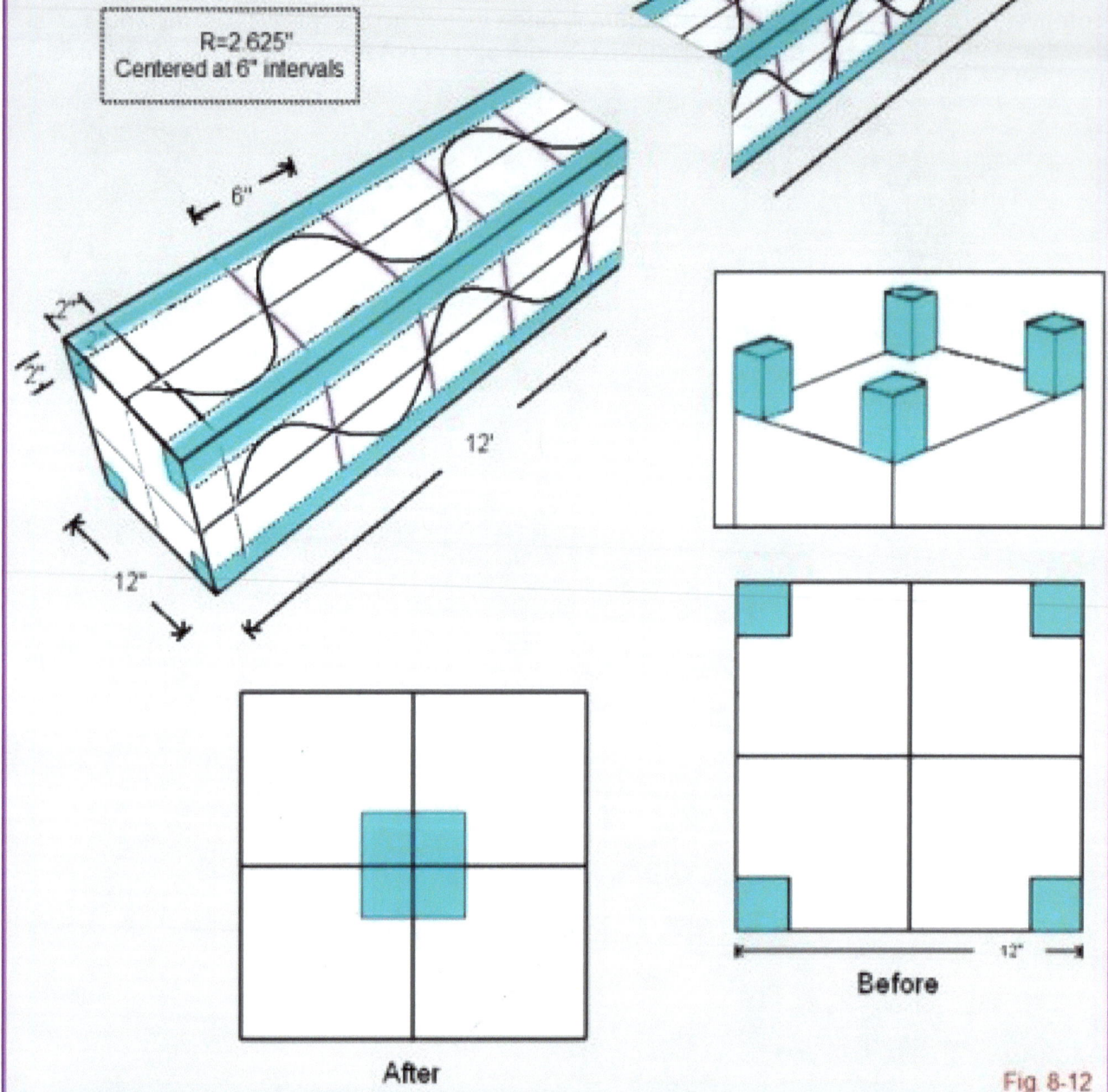

R=2.625"
Centered at 6" intervals

6"

12'

12"

12"

12'

After

Before

Fig. 8-12

I can help with the layout and cutting, but doweling and finishing would require advanced woodworking skills; both in the cutting and drying, and in the sanding and finishing.

Assuming you had a Falberg saw (ah-hem), you'd need to start with a true and squared beam of at least 6" x 6" sides of any length; with flat surfaces. I recommend a fine point black marker for laying out the cut pattern; pencil is fine too , but harder to see when the saw dust is flying around.

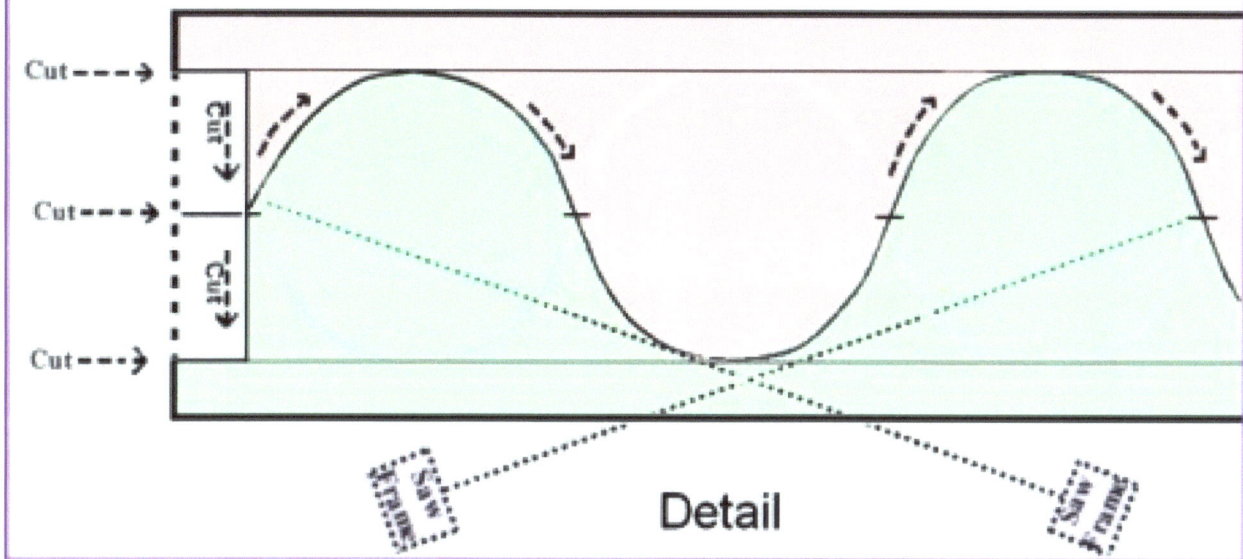

Fig 8-13

Cut ---→

Cut ---→

Cut ---→

Cut →

Cut →

Cut →

Saw Frame

Saw Frame

Detail

The exact dimensions will depend on where you end up after truing the beam so all I can do here is suggest the proportions based on my hypothetical 10" x 10" beam. The proportions themselves are negotiable and would depend how much stress you want the posts to carry. Also be advised that the concept of reciprocating radii need not be the "string of pearls" design shown here. Instead of the 3" radii I'm showing here, you could use two long arcs of a 10' radius. The result and the construction would be simpler, but it might lose some of its over-all appeal. Whatever pattern you choose, the closer you cut to the edge of the beam at any point, the weaker the post will be, structurally. I left a minimum 2" of free board along all four corner edges of my illustration. The finished construct will be as strong as its weakest link. If you leave a 2" border, the weakest link will be a 4" x 4". Considering the 4-ply nature of its construction, the 4" x 4" weakest link thus created should give you the structural integrity of a dimensional 4" x 4" , at least.

I used an aluminum disc of 6" diameter marked with a centerline. Laying the disc on my beam with the disc's center line in line with the beam's center line there was a 3" radius that ended 2" away from the edge of the beam. I marked the 2" border to ensure that I didn't cut outside that line; it also helps keep the disc centered. You could mark the border line at 3" and still use the 6" disc for a radius by simply shifting the disc off-center. If you do it this way you'll have more radius cuts per linear foot of post but that would be interesting too; try it. Send me some pictures. As long as you leave enough cross-cut-angle to avoid locking up the saw's frame against the side of the beam, it will work. I showed snapshots of the orientation of the saw in two of the cross-over positions to emphasize how important it is not to get locked into a 90* cross-cut. When you're crossing the centerline of the beam you have no wiggle room to get back to your scribe line, and once you get off-line the symmetry is lost. The trick is to draw your scribe line carefully, anticipating this, and then stay exactly on that line. If you DO get in trouble, you can force the blade over by "steering" it with the blade guides. (You'll have to re-adjust them again, soon as you get past the tight spot.)

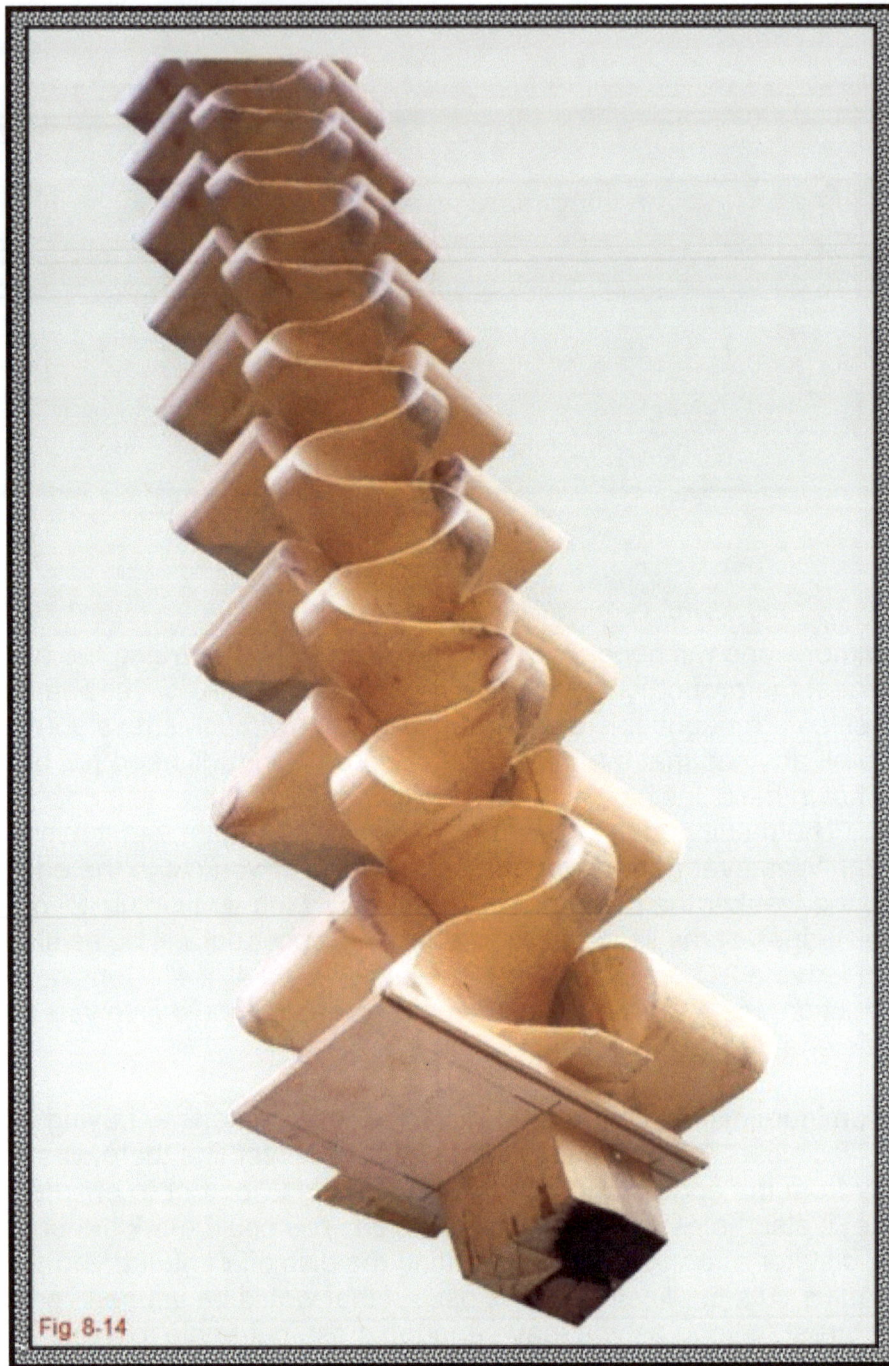

Fig. 8-14

You'd be well advised to label the end grain of the corner tabs that will form the base and cap ends. Once the clamps come off and this mess flops out on your saw horses you won't know where it all started. The full-lay-out illustration shows opposing wave-forms on the adjacent edges. You can draw them synchronously and it won't matter. Don't ask me why. I still don't understand it. You can reverse the pieces end-to end and re-assemble as you see fit, but no matter what you do you'll have two pieces that resemble a string of beads and two pieces that look like a slow-moving snake. The snakey pieces will form a plane along their ridged backs that will lay flat on your workbench. The snake-edge is fragile and needs to be hand sanded. I think the crux of Bayer's puzzle lies in that flat plane. I won't even try to describe it geometrically. It just is.

The work you're seeing here was done with prototype saws made in 2002 - 2003 which weren't nearly as easy to operate as the saws we make now. At the time of the original commissioned post, 3" was a really tight radius for the blade I was using, compounded by the added constraints of working at the limits of my available throat width. It was purely luck that the only blade I had to do the original post was very aggressive and it made the first such cut without problems. Subsequent blade purchases were universally 3/8"-2TPI Timberwolf blades. Without those legendary blades it's likely the Falberg Saw Co. would not have gotten off the ground. I didn't know it then, but that was the only blade in the world that could perform the deep-cutting rips and tight-radius turns I needed with such little blade tension as my saws were capable of at the time.

Even with an 18" throat width you have to *cheat* a little with your lay-out. It's not really cheating; magicians would call it creating an illusion. You can't cheat an honest man anyhow. If you were to actually link one semicircle directly to the next there would be a point, right at the intersections, where the saw's frame would have to be in the middle of the beam. That's physically impossible with a saw frame just eighteen inches behind the blade. The constraints thus imposed by this layout leave little margin for error. The three inch radius cuts were right up against the limits of the saw, the blade, and my ability to control it. Going slow, I could barely keep it on the scribe line around the curves and the frame bumped into the beam at every cross-over. You have to slant the cross-over just far enough to avoid a ninety degree cross-cut situation, and that will vary with the width of your throat. Therefore you have to leave three quarters inch between radii and draw straight-line links between them to avoid ever having a perpendicular cross-cut. When laying out the scribe line you should therefore pre-determine the maximum cross-cut angle given the width of your beam relative to the width of your throat. It's all very technical and it took me several hours to draw the first one. Hopefully the procedure described in this chapter will save you some frustration in that regard.

I didn't get any good pictures of the original, commissioned, *Totem* which was hand-sanded Doug Fir and finished with clear polyurethane. I made the gray post of Doug Fir also but it was finished without sanding; just brushed-on gray, oil-based paint. Sanding is optional here because the shape is the thing. The un-finished-wood post pictured above was done by one of my customers in pine and highlights the need for good hard wood as the snake-edge was chipped in several places. The exposed knots and grain were an unexpected bonus, however, and it looked fabulous when finished. Again, no pictures; sorry. The snake-edge, incidentally, formed a flat plane and is further testimony to the accuracy and consistency of the cuts. The snakey-edge, being the perpendicular intersection of two planes, is the most interesting feature of the sculpture in that it forms a third plane at forty-five degrees to the first two.

Can you imagine trying to maneuver a two hundred pound 10" x 10" x 12' beam through this series of snakey-cuts and keeping it flat on the scribe line? It should be obvious that infeed/outfeed tables don't apply to contour cuts of this magnitude. The level of strength, skill, and concentration required of the operator-s to do this is beyond my means; I can't speak for anyone else.

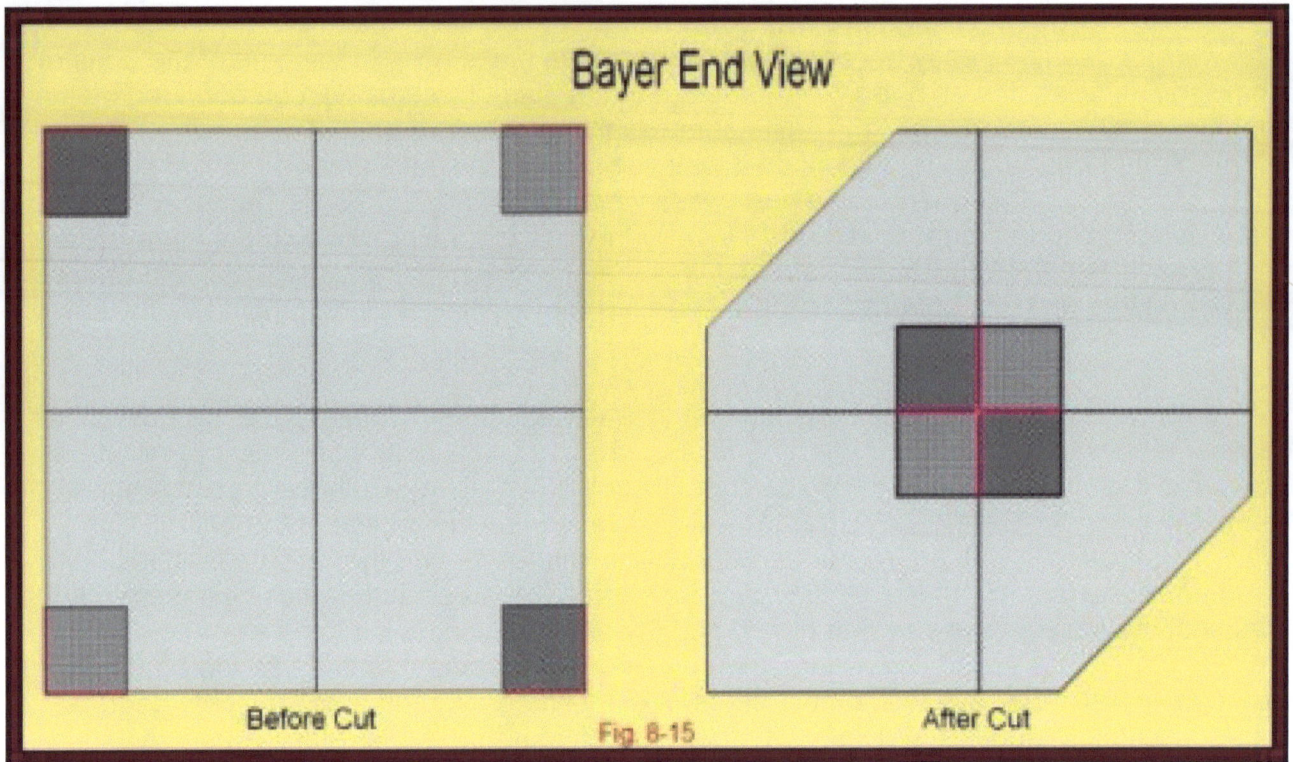

Bayer End View

Before Cut   Fig. 8-15   After Cut

I've never built a bandsaw box but I think they're just the coolest thing a guy can do with a bandsaw. I see them as an emerging trend in furniture design. When bandsaws evolve into the full-throated *light-sabers* they're destined to become, we'll see this concept explode into a totally new ***organic*** style of furniture design. The father of that design school is reading this now and understood the implications of the first seven chapters. I see an entire collection of household furniture crafted from laminated beams, in butcher-block fashion, into desks, dressers, tables, chairs, and beds. They will be cut with proprietary, home-made bandsaws with aluminum frames and a multitude of aluminum wheels. Jewelry boxes are just the beginning. What a great way to start, huh?

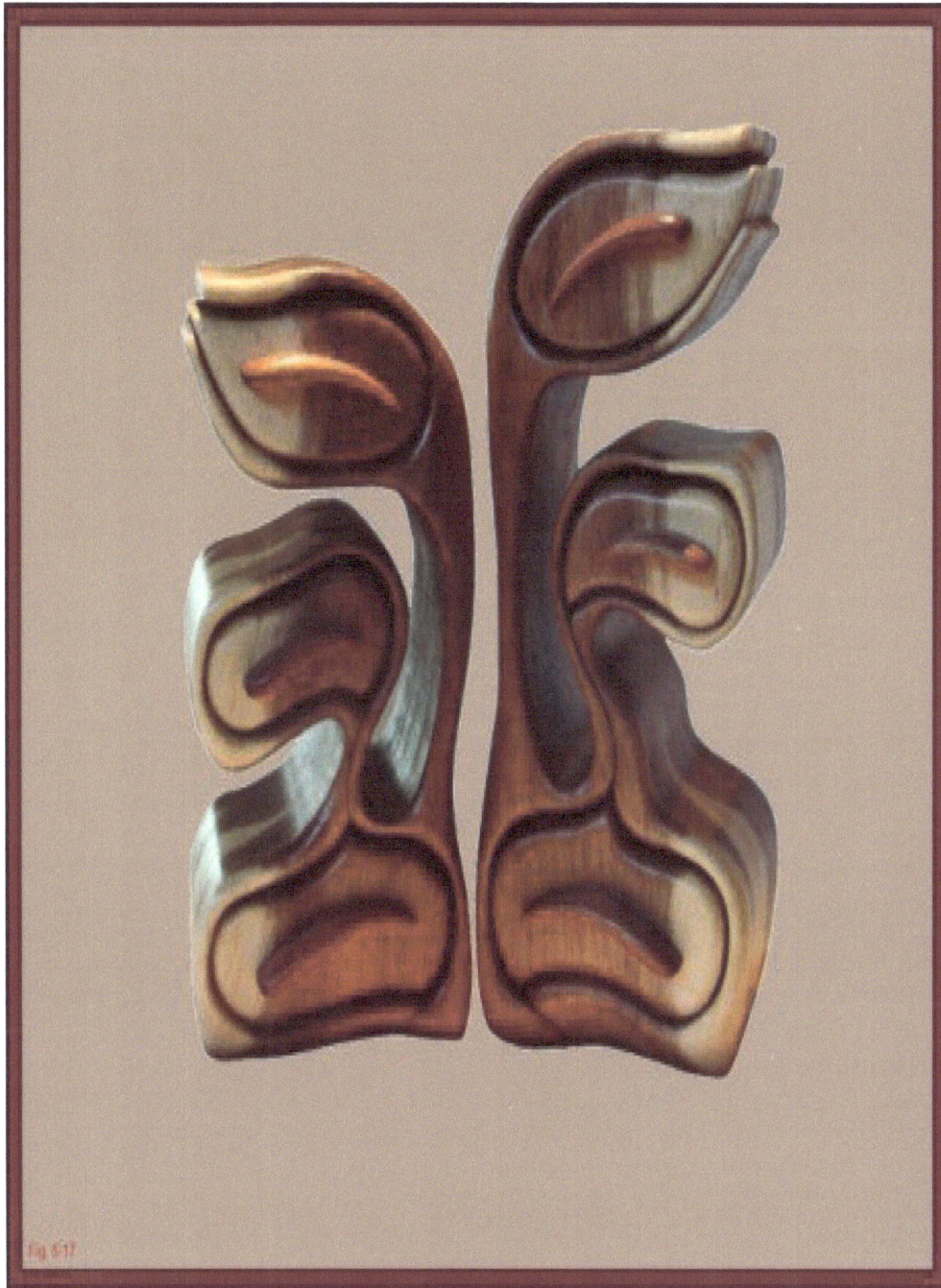

Fig. 8-17

Can you picture a **whole, forked, tree trunk** cut into drawers and shelfs as shown above.

I could write a whole other book about the safety aspects of operating a bandsaw but it wouldn't be much different than the books you've already read on the subject so what would be the point? It all comes down to using your eyes and ears and the space between them. Yada-yada ....attitude, respect the tools, keep a clean floor, alertness, and not picking them up by their cords.......yada-yada some more. The End.

Fig. 8-18

FALBERG SAW COMPANY

THINK SAFETY

My only contribution to the subject is to try, every time you pick up a tool, to **anticipate** all the crazy things that can go wrong and try to **imagine** the consequences. You can't avoid all risks but you can position yourself outside the line of fire: just in case. Contrary to what you may have heard: bandsaws are no more, nor less, safe than any other tool. Not to preach; it's your skin: but we need constant reminders. I print hundreds of these gory safety stickers and my customers always want more. They stick them on machines, helmets, mirrors, everywhere. You're free to copy and paste these anywhere you choose.

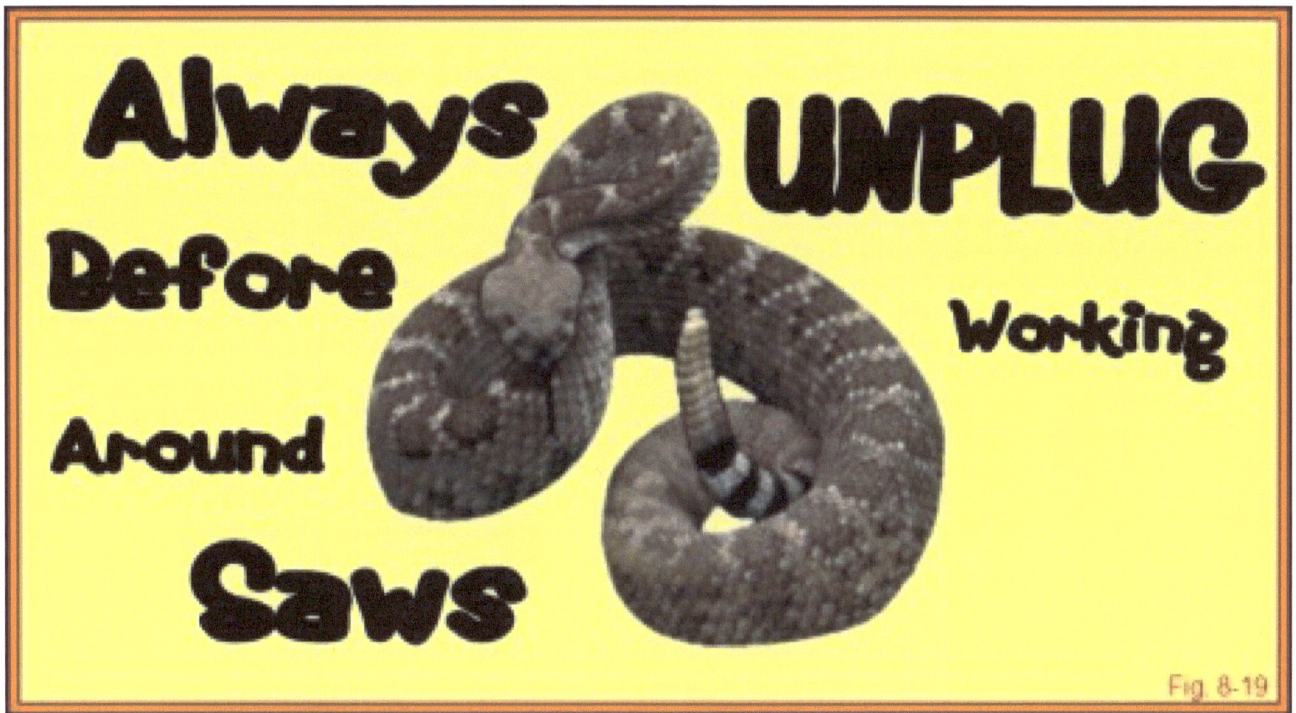

Always UNPLUG
Before
Working
Around
Saws

Fig. 8-19

Because Murphy, and his insane Law, will find a way to connect electrical power just as you stick your finger in there! .... It just does. Not often, but enough.

Cutting with a bandsaw usually ends with something dropping off. If the off-drop is big enough it can dent your toe or injure the lower half of your saw. The safety cover isn't only there to keep your fingers out; it keeps dropping drop-offs from dropping into high-speed wheel-wells. At 50 - 60 MPH a chunk of wood can do some serious hurt. The worse scenario, however, is when medium-sized slivers drop through the key hole and caught between the lower wheel and the blade; ka-chunking their way through the wheel housing. The blade can be thrown, at great velocity, right off the saw. It doesn't happen often, but when it does, it's spectacular. I wouldn't want to be in the way.

And speaking of dropping off; I hope all this hasn't put you to sleep. I'll shut up now.

The End

Acknowledgements:

Since my name is on the cover I'm forced to admit authorship and take responsibility for the contents herein. I'd rather blame it on the people who advised me, but I can't; they didn't write the final draft, nor is their name on it. Bad as my writing may, be it would certainly have been worse without the advice of friends and acquaintances too numerous to mention but too important to ignore entirely. You know who you are; so thank you (with a wink) all. Lest you come away thinking that my soul is twisted and my mind be evil please understand it was the work of my internal demon; not I. There's a voice in my head that shouts incessant irrelevancies to a do-wop chorus of "Do it! Do it! Do it!" Though my most trusted advisors advised against it, my contract with the voice was so favorable that the irreverent and silly content was left in. Neither this book nor the bandsaws it concerns are meant for children.

Foremost among notable people to thank is the mysterious Bob H., a **real** engineer and another builder of machines, for his extraordinary patience with my lack of math skills. Were it not for him, there wouldn't have been a section on blade stress testing and you'd still be laughing at my erroneous formulas. We have Bob H. to also thank for figure 2-03, the home-made stress tester. Mostly, he saved my scholastic reputation and standing in the intellectual community from ridicule.

Number-one son, Zubin Falberg, supplied the excellent eagles, flowers, and assorted sea bird photos as well as the clay work in Figure 3-08 and the glass globe in figure 6-13. Oh, and the praying mantis in figure 7-06. Nice shot. Love lost. He's a genuine artist, that boy.

Thanks to Gathering Wood (www.gatheringwood.com) for the saw mill in figure 1-10 and Miroslav Duchacek for the giraffe that happened by.

Thanks to Joe Grout for the picture of his spring tester set-up in figure 2-04 and his assistance in the blade tensioning discussion.

Thanks to Jody Card for not only his photos in figures 6-05, 6-06, 6-07, and 6-10; but for his priceless advisory support to the FSC itself. He's a fellow inventor and a Master Timber Framer.

Thanks to Bob Aquino for the tensioner photos in figures 2-09 and 2-11.

Thanks to Mark Conley for the old bandsaw photos in figures 6-14 and 6-15.

Thanks to John Toigo for the photos in figure 2-1and again in 2-12 which is shared by a photo from Darnell Hagen.

Thanks to Jerome Laux for his photo in figure 2-10 and a good bit of advice.

Thanks to Gary Katz for the Road Show photo of the Hull-Oakes Mill blade-zilla. If you haven't been to www.garymkatz.com ; you should. It's an education.

Thanks to Mark Rios for the photo of his shop-jig in figure 5-16.

Thanks to Ivan Mlinaric for his dog's commentary on my entertainment center in figure 7-19.

Thanks to Marvin Landis for his photo of the "Lotus Box" which he built from a design in Lois Ventura's *Building Beautiful Boxes With Your Band Saw* in figure 8-17.

Thanks to Matt Lippa at Folk Artisans for his photo of the mystery saw in figure 2-16.

About The Author

Bill Falberg spent most of his life in what we call the "trades" learning a variety of disciplines before settling down to general contracting. Semi-retired, he now lives in Grand Junction where he still pursues his life-long hobby of taking things apart to see what makes them tick; then pointing out to others how much better they would tick if they were made his way. It's an old habit that's led to several patent attempts, two of which were granted, only one of which was commercially successful - the balanced portable timber bandsaw. Although he points often, one is never sure what he's pointing at or why. His inscrutable eyes frighten children, but their penetrating intensity solve the incomprehensible mysteries of dark elemental forces, bringing order to the universe. Beyond that we only know that he's said by neighbors to be mild-mannered and somewhat reclusive.

Reading "Why My Band Saws Are So Cool And YOUR BAND SAW Sucks So Bad" will change your life forever! The reader learns herein: just how narrow the gap truly is between the veteran woodworker and the industrial tool designer; how to close that gap by learning a few cheap metal-working tricks; how few new tools are needed; and how to accomplish the transformation. Assuming the reader can design furniture and build cabinets, the book also assumes said woodworker is also capable of building tools, jigs, and entirely new machines out of aluminum with a minimum of additional cost and effort. It picks up where other books on the subject of band saws leave off and goes into much greater detail about the design elements incorporated into the building of dedicated industrial woodworking saws. In addition to the usual stuff you find in set-up and operation manuals this book goes on to teach the reader how to build a dedicated resaw that suits his individual needs precisely; how to build a custom band saw to cut 18" wide veneer for instance. The book teaches the reader how to build fences and saws far superior to anything currently available and in that regard one could make an entirely new career from these contents. If nothing else, the reader will come away with an entirely new understanding of how bandsaws work and how to better use them. Some may find humor in it.

Delving deeper into the theory of cutting solutions than other such books you'll find the contents bear serious study and for that reason I suggest buying the print version so it can be referred to frequently in the shop. The 168 page 8.5 x 11 printed book is done with ink-jet on glossy photo paper and comb-bound to lay flat. Unless it hits the New York *Best Sellers List* it will be home-made and each copy will be signed by me, personally. (If everybody in the World decides to buy it you'll have to sign them yourself.)

William H. Falberg

www.ingramcontent.com/pod-product-compliance
Lightning Source LLC
Chambersburg PA
CBHW041709210326
41598CB00007B/590